AQA
GCSE Science
CORE FOUNDATION REVISION BOOK

Mike Boyle, Philip Dobson
and Steve Witney

Hodder Murray
A MEMBER OF THE HODDER HEADLINE GROUP

Every effort has been made to trace all copyright holders, but if any have been inadvertently overlooked the Publishers will be pleased to make the necessary arrangements at the first opportunity.

Although every effort has been made to ensure that website addresses are correct at time of going to press, Hodder Murray cannot be held responsible for the content of any website mentioned in this book. It is sometimes possible to find a relocated web page by typing in the address of the home page for a website in the URL window of your browser.

Hodder Headline's policy is to use papers that are natural, renewable and recyclable products and made from wood grown in sustainable forests. The logging and manufacturing processes are expected to conform to the environmental regulations of the country of origin.

Orders: please contact Bookpoint Ltd, 130 Milton Park, Abingdon, Oxon OX14 4SB.
Telephone: (44) 01235 827720. Fax: (44) 01235 400454. Lines are open 9.00–5.00, Monday to Saturday, with a 24-hour message answering service. Visit our website at www.hoddereducation.co.uk

© Mike Boyle, Philip Dobson, Steve Witney 2007
First published in 2007 by
Hodder Murray, an imprint of Hodder Education,
a member of the Hodder Headline Group
338 Euston Road
London NW1 3BH

Impression number   5 4 3 2 1
Year   2011 2010 2009 2008 2007

All rights reserved. Apart from any use permitted under UK copyright law, no part of this publication may be reproduced or transmitted in any form or by any means, electronic or mechanical, including photocopying and recording, or held within any information storage and retrieval system, without permission in writing from the publisher or under licence from the Copyright Licensing Agency Limited. Further details of such licences (for reprographic reproduction) may be obtained from the Copyright Licensing Agency Limited, Saffron House, 6–10 Kirby Street, London EC1N 8TS.

Cover photos Science Photo Library: dragonfly, Andy Harmer; house, Ted Kinsman; limestone, Alfred Pasieka
Illustrations by Barking Dog Art
Typeset in Times New Roman 11.5pt by Fakenham Photosetting Ltd, Fakenham, Norfolk
Printed in Italy

A catalogue record for this title is available from the British Library.

ISBN-13: 978 0340 914 23 6

Contents

2 Biology

2 B1a Human biology
- 2 Nervous system and reflexes
- 4 Hormones, the menstrual cycle and fertility
- 6 Controlling internal conditions
- 8 Drugs: good and bad
- 10 Tobacco and cannabis
- 12 Diet and exercise
- 14 Problems with 'bad' diets
- 16 Fighting disease
- 18 The fight against disease – then and now

20 B1b Evolution and environment
- 20 Adapt and survive
- 22 Populations and competition
- 24 Variation and inheritance
- 26 Reproduction and cloning
- 28 Genetic engineering
- 30 The fossil record
- 32 How evolution happens
- 34 How do humans affect the environment?
- 36 Global warming
- 38 Sustainable development

40 Chemistry

40 C1a Products from rocks
- 40 Atoms, elements and the Periodic Table
- 42 Reactions, formulae and balanced equations
- 44 Products from limestone
- 46 Extracting metals
- 48 Quarrying, mining and recycling
- 50 Using metals and alloys
- 52 Crude oil and fuels
- 54 Burning fuels
- 56 Cleaner fuels

58 C1b Oils, Earth and atmosphere
- 58 Cracking crude oil
- 60 Making ethanol
- 62 Making polymers
- 64 Waste-disposal problems
- 66 Vegetable oils and fuels
- 68 Food additives and emulsifiers
- 70 The Earth and continental drift
- 72 Plate tectonics
- 74 Gases in the atmosphere
- 76 Theories about the atmosphere

78 Physics

78 P1a Energy and electricity
- 78 Thermal radiation
- 80 Conduction and convection
- 82 Reducing rates of heat transfer
- 84 Energy efficiency
- 86 Electrical power and energy costs
- 88 Generating electricity and the National Grid
- 90 Renewable energy resources
- 92 Comparing energy resources

94 P1b Radiation and the Universe
- 94 Waves and electromagnetic waves
- 96 Uses of electromagnetic waves
- 98 Hazards of electromagnetic waves
- 100 Radioactivity
- 102 Uses of radioactivity
- 104 Stars and telescopes
- 106 Expanding Universe and 'big bang'

108 Answers

113 Index

B1a Human biology

Nervous system and reflexes

The nervous system

- The nervous system allows you to **detect** what's going on inside and outside your body, and to **respond**.
- It does this by detecting **stimuli**, which are changes that we can detect, such as light, sound, movement or heat.
- All this information is sent into your **central nervous system** (**CNS**). The CNS is made up of your **brain** and **spinal cord**.
- **Receptors** are cells that can detect a stimulus. There are many different stimuli, so there are many different receptors.
- **Sense organs**, such as the eyes and ears, contain a lot of receptor cells of one type. The receptors in the eyes are sensitive to light, and so on.
- Receptor cells send this information along specialised cells called **neurones**, or **nerve cells**, which are like tiny wires.

Most of this sensory information goes to our **brain**, which **processes** the information and **coordinates** a response.

The organ that responds is called the **effector**. This could be a **muscle** or a **gland** that responds by releasing a chemical such as a **hormone**.

The senses

There are five senses. They are:

Reflexes

A **reflex** is a quick response. Examples of reflexes include:

- blinking when something comes near to your eye;
- jerking your hand away from a hot or sharp object;
- the knee jerk (when tapped just under the kneecap).

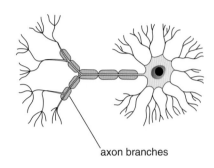

A single neurone or nerve cell

■ Neurones (or nerve cells) can be very long and can transmit information from one part of the body to another

exam tip

★ In exams, many students confuse sense organs with receptors – 'the skin is a receptor' would get no marks. The skin is a **sense organ** that contains many different **receptor cells**.

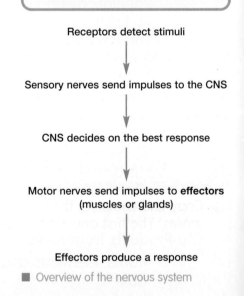

■ Overview of the nervous system

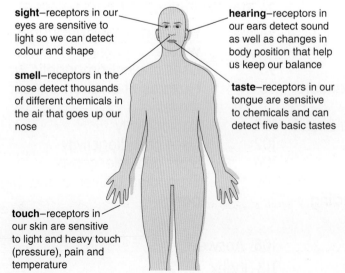

sight—receptors in our eyes are sensitive to light so we can detect colour and shape

hearing—receptors in our ears detect sound as well as changes in body position that help us keep our balance

smell—receptors in the nose detect thousands of different chemicals in the air that goes up our nose

taste—receptors in our tongue are sensitive to chemicals and can detect five basic tastes

touch—receptors in our skin are sensitive to light and heavy touch (pressure), pain and temperature

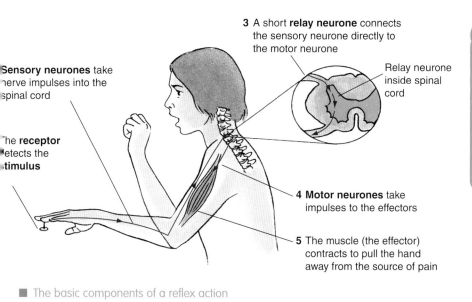

- **Sensory neurones** take nerve impulses into the spinal cord
- The **receptor** detects the **stimulus**
- 3 A short **relay neurone** connects the sensory neurone directly to the motor neurone
- Relay neurone inside spinal cord
- 4 **Motor neurones** take impulses to the effectors
- 5 The muscle (the effector) contracts to pull the hand away from the source of pain

■ The basic components of a reflex action

exam tip

★ Always talk about nerve **impulses**, not messages. A nerve impulse carries information, but it's the brain's job to make sense of the tiny electrical blips and 'decode' the message.

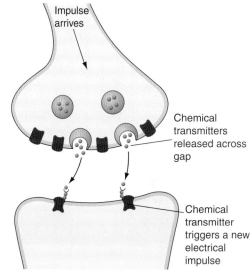

■ A synapse

The key features of reflexes are:

- they are fast;
- they are automatic – they do not involve the conscious control of the brain, so you can't stop them even if you want to;
- they are always the same – the **same stimulus** leads to the **same response**.

Synapses

A **synapse** is a junction between two neurones. Synapses occur between relay neurones and sensory neurones in a reflex action.

When a nerve impulse reaches a synapse, a **transmitter chemical** is released. This passes (diffuses) across the gap and sets up an impulse in the next neurone.

check your understanding

1. Copy and complete the table for each of the five senses. The first one is done for you. (12 marks)

Sense	Sense organ	Receptor cells sensitive to
Sight	The eye	Light

2. a) What is a neurone and what does it do? (2 marks)
   b) State the difference between sensory and motor nerves. (2 marks)

3. Use the following words to complete the paragraph below. (3 marks)

   brain   muscle   motor   sensory   stimulus

   To make us aware of a ____1____ such as a loud noise, impulses are sent along ____2____ neurones to the ____3____.

FOUNDATION

B1a Human biology

Hormones, the menstrual cycle and fertility

Hormones are chemical signals

- **Hormones** are compounds made and released by **glands**, such as the ovaries, testes and pituitary gland.
- They always travel in the blood.
- They have an effect on **target cells**. These cells may be in one particular organ, or scattered throughout the body.

The menstrual cycle

The **menstrual cycle** is the woman's monthly reproductive cycle. Every month an egg, or **ovum**, develops in one of a woman's **ovaries** and then released. The release of an egg is called **ovulation**.

There are three hormones involved – **FSH**, **oestrogen** and **LH**.

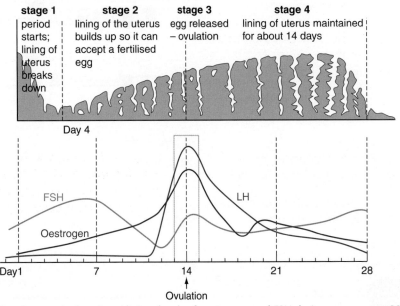

■ The changes in the uterus lining, levels of oestrogen and FSH during an average 28-day menstrual cycle

exam tips

Adrenaline is the hormone that we release when we are scared, anxious or nervous about something. You are not expected to know about adrenaline, but if you were told in an exam question that when your body releases adrenaline:

★ your heart beats faster;

★ you breathe more deeply;

you would be able to state that the target cells for adrenaline are in the heart and the breathing muscles.

★ If the exam asks you to 'use the information', you **must** base your answer on the information provided.

★ An ethical objection to IVF treatment might be the opinion that an embryo is a human being, with rights, from the moment the sperm meets the egg. IVF creates many embryos that will never be used, and this could be seen as a waste of human life.

Hormone	Full name	Where it's made	What it does	When
FSH	Follicle-stimulating hormone	Pituitary gland	Causes the development of an egg inside a follicle	Day 1 (when the period starts) to about day 13
			Causes the ovary to release oestrogen	
Oestrogen	Oestrogen	Ovary	Causes uterus lining to thicken	Day 5 (after period) to day 28
			Stops production of more FSH	About day 12/13
			Causes the pituitary gland to produce LH	Day 13/14
LH	Luteinising hormone	Pituitary gland	Causes ovulation	About day 14

FOUNDATION

● Hormones can be used to control fertility

Hormones control the development of eggs and ovulation, so we can use hormones as **contraceptives** to prevent pregnancy.

The contraceptive pill ('**the pill**') contains oestrogen. High levels of oestrogen stop the production of FSH. No FSH, no egg develops, no ovulation, no baby. Sometimes **infertility** is caused by **low levels of FSH**, so that eggs do not mature and cannot be released.

You might be given some information about the pill and asked to describe its advantages and disadvantages, as in the table.

We can also use hormones in **fertility treatment** to help couples conceive a baby. In **IVF** (*in vitro* **fertilisation**), the woman is given FSH so that her ovaries **produce more eggs** than normal, and LH to **stimulate the release** of these eggs. The eggs are collected and fertilised by sperm *in vitro* (in a test tube). The fertilised eggs develop into embryos and are put into the uterus.

● Risks and ethical problems of fertility treatment

- Fertility treatment may not work – trying again and again takes time, and may be stressful and expensive.
- It may cause the ovaries to make too many eggs, resulting in **multiple births** (triplets, quads, etc.).
- Ethics is about what people think is right or wrong, such as 'is it right to do this?' Making decisions about using science is not always about what **can** we do, it's what **should** we do.

Advantages
It is very effective. 99% of women on the pill do not get pregnant – but some do
It may make periods lighter and less painful
It doesn't interrupt love-making, in the way that condoms do
Disadvantages
It gives no protection against sexually transmitted diseases such as HIV and chlamydia
It may have side effects, such as headaches, feeling sick, irregular periods and water retention
Women must remember to take it every day
It is against the teaching of some religions

exam tip
★ If the exam asks you to 'use the information', you must base your answer on the information provided.

check your understanding

4 Which of these hormones is **not** involved in controlling fertility? *(1 mark)*
  A oestrogen   B luteinising hormone (LH)
  C follicle-stimulating hormone (FSH)   D testosterone

5 Which of these organs is **not** involved in the menstrual cycle? *(1 mark)*
  A ovaries   B stomach   C uterus   D pituitary gland

6 Use the following words to complete the paragraph below. *(3 marks)*
  nerves   blood   gland   target organ   hormone   muscle

  Oestrogen is an example of a ____1____. It is released by the ovaries and travels in the ____2____. Its ____3____ is the uterus, where oestrogen causes the lining to thicken.

7 Using the information on this page, describe the problems that may occur due to the use of hormones to assist fertility, including IVF. *(4 marks)*

FOUNDATION

B1a Human biology

Controlling internal conditions

Homeostasis is all about keeping the conditions inside your body as stable as possible. This is important because if your cells are not kept in just the right conditions, they can be harmed.

● Four examples of homeostasis

1 Control of water content
We don't want to become dehydrated, or over-hydrated, so the water that goes into our bodies must be balanced by what comes out.

We can lose water in:

- sweat;
- urine;
- our breath;
- faeces (we lose a lot more water when we have diarrhoea).

This water loss must be replaced by our food and drink.

2 Control of ion content
Important ions include sodium, chloride (sodium chloride = common table salt), potassium and calcium. These ions are sometimes called **electrolytes**.

We gain ions from our food and drink. We lose ions when we urinate and when we sweat.

3 Control of body temperature
The temperature in the centre of your body stays very constant whatever the temperature of the surroundings.

Your brain can detect the temperature of the blood flowing though it, and make responses that increase or decrease body temperature as needed e.g. sweating or shivering.

4 Control of blood glucose
Glucose in the blood must be kept within certain limits. People who cannot control their blood glucose levels are **diabetic**.

Control of blood glucose levels is an example of how hormones work.

Receptor cells in the pancreas detect increased or decreased levels of glucose.

The cells respond by making and secreting hormones.

These hormones circulate in the blood and make cells take in or release more glucose from the blood until the level is back to normal.

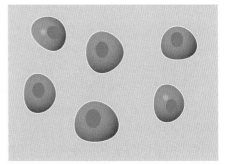

■ All the cells in your body get their food and oxygen from tissue fluid surrounding the cells. They also put waste into it. The blood circulation keeps this tissue fluid 'fresh' so that the cell contents can be maintained.

Evaluating sports drinks

When we exercise, our muscles use glucose as a source of energy, and we lose water and ions in sweat. All the glucose, salt and water need to be replaced. Sports drinks contain water, glucose and ions.

Imagine you are investigating the claim:

> **'Our sports drink keeps you going for 33% longer'**

You could design your own **investigation** to see if the findings can be repeated. The conclusion is **valid** if the athlete's performance is affected only by the sports drink, and not by **other factors**.

Here are some questions you might like to think about when planning this investigation.

- What would you measure? What you measure is the **dependent variable**. The **independent variable** (the one you change) is drinking the sports drink, but what would you compare it against – what would be your **control variable**?
- How many athletes would you give the sports drink to? The sample must be large enough to make the investigation **reliable**.
- How would you choose your sample of athletes? There would be **bias** if most were very fit.
- How many times would you repeat the experiment? You need enough data to make your investigation **reliable**.
- You could investigate if you get the same effect from simply drinking water with a little salt and glucose dissolved in it.

exam tip

★ To **evaluate** a claim, use the information provided to think about the evidence **for** and **against**, and draw a conclusion.

■ Sports drinks are designed to replace the ions we lose

exam tip

★ When we say that data are **reliable**, it means that they can be trusted because other measurements (by another person, or another technique) have given the same result.

check your understanding

8 a) State why it is necessary for the body to maintain the level of glucose in the blood. *(1 mark)*
b) List the processes by which water is lost from the body. *(3 marks)*

9 Use the following words to complete the paragraph below. *(3 marks)*

dehydrated   sodium   water   sweat   glucose   energy

In order to keep cool, marathon runners have to sweat. This works well, but the problem is that they lose a lot of ____1____, and this can make them ____2____. At the same time, they can lose important ions such as ____3____. Throughout the race, they have to drink large amounts to replace what they have lost.

10 Which of the following is **not** controlled by internal mechanisms in the body? *(1 mark)*

A Temperature.
B The amount of glucose we eat.
C Water output.
D Concentration of ions in cells.

FOUNDATION

B1a Human biology

Drugs: good and bad

Drugs are compounds that affect your body's chemistry in some way. There are two main types of drugs.

- **Medical drugs** are medicines to treat a particular illness. They could still be used illegally, and could harm the body.
- **Recreational drugs** are taken for their short-term effects. Some, such as cocaine, heroin and LSD, are illegal.

Why do people take recreational drugs?

It's usually because people like their **short-term effects**:

- relaxation;
- stress relief – an escape from everyday worries;
- stimulation – some drugs (e.g. amphetamines or 'speed') combat tiredness, so that people can work longer or 'party all night';
- inspiration – some people claim they are more creative while under the influence of drugs;
- pain relief – many multiple sclerosis sufferers claim that cannabis gives relief from muscular pain.

However, these drugs can lead to serious problems including **addiction**, which means that you keep wanting more. Heroin and cocaine are highly addictive. Alcohol and cigarettes are also addictive, and users can suffer **withdrawal symptoms** if they try to give up.

Alcohol – legal, but dangerous

Alcohol reduces the activity of the nervous system. This means:

- slower brain function – you think more slowly;
- **reactions** are **slower**;
- **co-ordination** and **judgement** are affected, which is why drinking-and-driving is illegal;
- loss of **self control**, so people do things they wouldn't do when sober, such as saying 'yes' to sex.

Excess alcohol in one session (binge drinking) can cause **dehydration** (the main cause of hangovers), **unconsciousness** and even a **coma**. **Long-term** alcohol abuse can result in liver damage (**cirrhosis**) and **brain damage**.

Because so many people drink alcohol, the cost of treating problems, the accidents caused by drink-drivers, the number of working days lost, and the **social costs** to families can be very high.

FOUNDATION

Testing and trialling medical drugs

Before they can be used on patients, medical drugs have to pass tests to make sure they work and are safe to use (diagram right).

For the data in a clinical trial to be **reliable**, the study must be large, **controlled** and able to be **reproduced** by other scientists. As a control, some patients are given a **placebo** that does not contain the drug.

Thalidomide – a drug that wasn't properly tested. In the 1950s, thalidomide was tested as a sleeping pill. Trials showed that it was also very effective against **morning sickness**, so it was given to **pregnant women**. The drug had not been tested on pregnant women.

Disastrously, the drug affected the **development of the fetus** (unborn baby), so that many children were born with very short and deformed arms or legs. The drug was banned in 1962, and better testing procedures, like those in the flowchart, were developed for new drugs.

Testing new products on animals is controversial because it may involve suffering and pain. You need to be able to offer opinions on both sides of the argument – there are some examples in the table.

1. Animals, plants or fungi can be tested to see if they contain substances that may be useful as new **painkillers** or **antibiotics**.
   ↓
2. Laboratory tests on **human tissue samples** – sometimes this gives good results, but it can't show the effects on the whole body. For instance, you can't test the effectiveness of a sleeping pill on a sample of tissue.
   ↓
3. Test the drug for toxicity and effectiveness on **live animals** – this usually has to be a mammal, because humans are mammals.
   ↓
4. **Clinical trials** – the drugs are tested on human volunteers to see if they are effective and if there are any **side effects**. The trials start with a low dose.

■ How new drugs are developed

Against	For
Testing drugs on animals is morally wrong because the animals suffer	The suffering of animals is unfortunate, but the benefits are more important – the lives of many humans and animals are made better by the medicines that are proved to be safe
Drugs don't need to be tested on cats, dogs and monkeys – scientists could use animals that have no sense of pain, such as jellyfish, or cockroaches	These animals are so different from mammals that the results would not be valid – tests are needed on mammals

check your understanding

⑪ a) What does addiction mean? (2 marks)
   b) Name two illegal drugs that are addictive. (2 marks)
   c) Name two legal drugs that are addictive. (2 marks)

⑫ Use the following words to complete the paragraph below. (4 marks)

   clinical trials   human tissue   side effects   live animals

   A new painkiller must be tested to make sure it has no ____1____. First, the drug is tested on samples of ____2____. Then it has to be tested on whole organisms, which involves using ____3____ to check for toxicity. Finally, the drug is tested in ____4____.

⑬ There are high social costs of alcohol abuse. Which one of the following is **not** a possible cost of alcohol abuse? (1 mark)

   A Days off work.   B Medical treatment.
   C Profits made by off-licences and pubs.   D Divorce.

FOUNDATION 9

B1a Human biology

Tobacco and cannabis

Tobacco is legal in the UK for over-16s, while cannabis is illegal. Some people want to make cannabis legal, but this is controversial.

● What's in tobacco smoke?

- **Nicotine** – this relaxes smokers. It is **highly addictive**.
- **Tar** – contains many chemicals, some of which are **carcinogenic** (cancer-causing).
- **Carbon monoxide** – a gas that sticks to haemoglobin, preventing the blood from carrying as much oxygen as it should. If pregnant women smoke, the baby can be born underweight.

● There is a well established link between smoking and lung cancer

There is a **causal link** between **smoking** and **lung cancer**, which means that smoking causes lung cancer. Cancer Research UK estimates that 90% of lung cancer cases are due to smoking.

However, you can **never prove** that smoking caused one smoker's lung cancer, we can only say that it is **highly probable**.

● Evaluating ways to stop smoking

Suppose you were asked to compare methods to see which is best. You could listen to someone's **opinion**. The problem is that this is not very **reliable**, because it is based on a sample of one or a few people.

You could see what the manufacturers of nicotine patches claimed, but their claims might be **biased**. (They're not likely to say that their patches don't work, or are no better than anybody else's!) They might claim to have **evidence**, but their evidence may not be **credible** because it wasn't gathered scientifically.

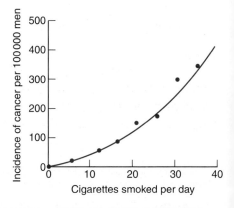

■ The **correlation** between smoking and deaths from lung cancer was noticed in the 1940s. However, smoking as a **cause** was not accepted until more **evidence** came from interviews with many lung cancer patients. The **statistics** showed that nearly all were heavy smokers.

You could **evaluate** effectiveness by looking at data on the number of smokers using different methods. To draw a conclusion, see if there is a pattern. Remember to think about:

- Is the **sample size** large enough?
- Was the method **valid** – comparing only one factor?
- Could there be **other factors** – such as age or workplace environment of the smokers – that the survey didn't take into account?

> **exam tip**
>
> ★ When you are asked to **evaluate** it means judge. Read the information provided and think about it. Look at the evidence **for** and **against**.

What is cannabis?

Cannabis is a plant – its leaves are dried and smoked, producing a feeling of relaxation and mellowness. Many people think it is relatively harmless, but it is still **illegal**.

Objections to cannabis include:

- many studies have linked cannabis use with **mental illness** such as **schizophrenia** – this could be due to chemicals in cannabis that affect the brain;
- people say that cannabis use could lead to 'harder' (more addictive) drugs – there is some evidence that most users of hard drugs have tried cannabis first.

If you were asked to evaluate the **link between cannabis and hard drugs**, use the information provided to consider the evidence for and against. Be careful to separate **opinions** (which are affected by social and moral judgements) from **evidence**, such as long-term studies on large groups of people, including a control group.

> **check your understanding**
>
> ⑭ In the UK in 2004/05, 26% of men and 23% of women were cigarette smokers, compared with the early 1970s when around 50% of men and 40% of women smoked.
>   a) Explain why it can be difficult to give up smoking. *(2 marks)*
>   b) Suggest why fewer people smoke now compared with the 1970s. *(1 mark)*
>
> Deaths due to lung cancer have also decreased since the 1970s.
>   c) Apart from the fall in the number of smokers, suggest one reason why the numbers of people dying from lung cancer have fallen. *(1 mark)*
>   d) State one other advantage for women of stopping smoking. *(1 mark)*
>
> ⑮ A scientist notices that the number of cases of skin cancer is highest in the areas of highest ice cream sales. This is called a correlation.
>   a) From this information, is it correct to say that eating ice cream causes skin cancer? Give a reason for your answer. *(2 marks)*
>   b) Suggest one other factor that might explain the high number of skin cancer cases. *(1 mark)*

FOUNDATION

B1a   Human biology

Diet and exercise

The saying 'you are what you eat' is true – our bodies are made up from the food we have eaten. However, if our food contains more energy than we use, we will store the extra as fat, and this can be a big problem.

● Balanced diet – the right amount of energy and a balance of different nutrients

- **Carbohydrates** (such as sugars and starch) for energy.
- **Fats** to provide **insulation** and energy.
- **Proteins** for growth, cell repair and cell replacement.
- Small amounts of **minerals** and **vitamins**.
- **Fibre** to allow the gut to push food along efficiently. Most fibre is plant material that we can't digest.

■ If you want to stay healthy, five words of advice are: eat plants, exercise, don't smoke

People who are starving are **undernourished**. Children fail to grow and develop normally. Women may have **irregular periods**. The lack of food can cause a **reduced resistance to infection** – people are less able to fight off disease.

People who don't eat a balanced diet are **malnourished**, and their health will suffer in some way. This could include being overweight or too thin.

A lack of a particular vitamin or mineral will cause a **deficiency disease**. For example, a lack of **vitamin D** can cause **rickets** – a condition in which the bones don't develop properly.

● How much energy do we need?

- All of us need a certain amount of energy, which we get from our food.
- Proteins, carbohydrates and fats **all contain energy**.
- This energy is used for many different chemical reactions in the body, and to move muscles and keep you warm.
- **Exercise** (moving muscles) increases the amount of energy used by the body.

● Balancing energy in food and energy used

The only way to lose weight is to change the energy input/output balance, which means reducing the energy from food and being more active.

FOUNDATION

- **Metabolism** is the general term for the chemical reactions inside your body.
- **Metabolic rate** refers to the speed or rate of these reactions.
- People with a **high metabolic rate** will tend to be thin, and can eat a lot of food without putting on weight; people with a **low metabolic rate** tend to be overweight, and a higher proportion of their food will be stored as fat.

The amount of energy you need from food depends on:

- **your age** – young people usually have a higher metabolic rate than older people;
- **your sex** – boys usually have a higher proportion of muscle to fat than girls do, and muscle has a higher metabolic rate;
- **your genes** – some people **inherit** a tendency to have a high or low metabolic rate, or a certain muscle-to-fat ratio;
- and, of course, it depends on how much exercise you do – **active muscles** have a very **high metabolic rate**, and when we exercise our metabolic rate **stays high** for quite a while afterwards.

Exercise is a great way to control weight because it uses up energy from food **and** it raises metabolic rate.

Activity	Energy used in kJ/min
Standing, cooking	9
Sitting, watching TV	5
Walking briskly	15
Climbing stairs	28
Swimming	35
Dancing	19
Jogging	27

check your understanding

16 Use the following words to complete the paragraph below. *(3 marks)*

glucose   metabolic rate   energy requirements

During the Tour de France race, cyclists are on their bikes for long periods and for many days. This means their _____1_____ will be very high for long periods. Their daily energy needs are about twice that of the average male of the same age. The cyclists take regular drinks containing _____2_____, our body's immediate source of energy. Being so active, these athletes will have a high _____3_____.

17 Compared with most males of the same age, do cyclists usually: *(1 mark)*

   A eat more and have a low metabolic rate?
   B eat less and have a low metabolic rate?
   C eat more and have and a high metabolic rate?
   D eat less and have a high metabolic rate?

18 Explain the difference between malnourished and undernourished. *(2 marks)*

19 People who move to hotter countries often find that they put on weight. Suggest why you need less food in a warm climate than in a cold one. *(1 mark)*

20 The table above shows the energy used by an average 25-year-old female office worker.
   a) Which activity uses the most energy? *(1 mark)*
   b) Explain why a professional dancer has a higher energy requirement than an office worker who goes to a dance class after work. *(1 mark)*

B1a Human biology

Problems with 'bad' diets

● The problems of too much food

In the developed world, an increasing number of people are overweight or **obese**, which means grossly overweight.

Obesity can cause health problems:

- **heart disease** (see below);
- **high blood pressure** – the heart has to work harder to pump blood around all that extra body tissue;
- extra strain on the joints, which may cause **arthritis**;
- **diabetes** – a disease with many possible complications.

● Too much salt is also a problem

For some people, too much salt can cause high blood pressure. These people need to watch how much salt they eat. It's easy to exceed the recommended daily dose of 6 g of salt per day.

■ Most foods contain some salt to start with, *more* salt is added when food is processed, and many of us add *even more* salt at the table

● High cholesterol levels can lead to cardiovascular disease

Although cholesterol is a natural substance needed by cells, too much cholesterol in the blood increases the risk of **coronary heart disease** and diseases of the **blood vessels**. These diseases are caused when arteries become narrower due to a build-up of **fatty deposits** inside them. When arteries are blocked, not enough oxygen reaches the heart muscle. A **heart attack** may result.

Cholesterol levels in the blood are affected by:

- the amount produced by the **liver**;
- the amount in the foods we eat – animal fats are especially high in cholesterol;
- genetic factors – some unlucky people have a high level of cholesterol, and these people have to watch what they eat.

What's the link between high cholesterol and lipoproteins?

Cholesterol is a type of fat, so it doesn't dissolve in water. So to be transported in the blood, cholesterol has to be attached to other molecules, called **lipoproteins**.

There are two types of lipoprotein: **high-density lipoproteins** (**HDLs**) and **low-density lipoproteins** (**LDLs**). LDLs are 'bad' – they carry cholesterol from the liver and take it to other cells in the body. Having a lot of LDLs can lead to build-up of cholesterol deposits in arteries. HDLs are 'good' – they 'mop up' cholesterol and take it in the opposite direction, from body cells to the liver.

It's all about balance. You want a high ratio of HDL to LDL in your blood.

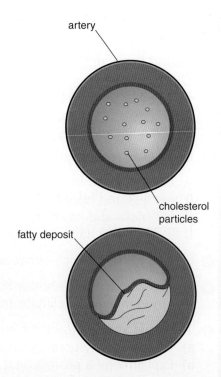

■ A healthy artery (top); the one below shows signs of coronary heart disease. The gap is narrower and blood flow will be reduced.

Saturated and unsaturated fats

Saturated fats **raise** cholesterol levels.

Unsaturated fats **lower** blood cholesterol. They also improve the HDL/LDL balance.

All fats should be eaten in moderation, but the type of fat in the diet is also important. Saturated fats are common in **animal products** – meat, full-fat milk, cream and butter. Unsaturated fats are found in fish oils and vegetable oils. Processed food often contains a high proportion of saturated fat.

Statins can lower cholesterol

Statins are new drugs that can lower cholesterol levels. They stop the liver from making LDLs. They could be life-saving drugs, but are there any side effects?

To find out, the British government carried out an **independent survey**, which means they had no connection with the manufacturers. The sample was large – over 6000 patients were tested, so the results were **reliable**.

Studies such as these need **two groups**: one group that has taken statins and another **control** group, as identical as possible, that has not.

They found that people who had taken statins had significantly lower cholesterol levels than the control group.

check your understanding

㉑ Use the following words to complete the paragraph below. *(5 marks)*

        statins   obese   side effects   saturated fat   heart disease

My aunt says that my uncle Dave is a walking heart attack. He's 6 feet tall but weighs 22 stone, so he's ____1____. Also, he's a long-distance lorry driver and he eats in motorway services and transport cafés. He smokes, and his favourite food is steak, egg and chips. His diet is high in ____2____. He complains of chest pains, and the doctor said that he has the symptoms of ____3____. They did a blood test and his cholesterol level was high. The prescribed him some ____4____, but he is worried about taking them as there could be ____5____.

㉒ The term 'risk factor' means that you are more likely to suffer from a particular disease. List two risk factors for heart disease. *(2 marks)*

㉓ a) List two foods that cause high cholesterol levels. *(2 marks)*
    b) Explain why a high cholesterol level can lead to coronary heart disease. *(1 mark)*
    c) Suggest how someone at risk for coronary heart disease could reduce their blood cholesterol level. *(1 mark)*

FOUNDATION

B1a Human biology

Fighting disease

There are two types of disease.

- **Infectious** – diseases that you can catch. These are caused by **pathogens**, which are **disease-causing microorganisms**. The main ones are **bacteria** and **viruses**.
- **Non-infectious** – diseases you can't catch. These are caused by other factors, such as lifestyle (diet, smoking, etc.), your genes, or simply getting older.

● Bacteria are very small, but a major cause of disease

Bacteria are tiny, single-celled organisms. Human cells are about 1000 times bigger than bacterial cells.

Only a few types of bacteria cause disease.

Bacteria cause disease when they multiply (**reproduce**) inside our bodies.

● Viruses make bacteria look big

Viruses are extremely small – even some molecules are bigger than viruses!

Viruses cannot reproduce on their own. They reproduce by getting inside your cells and instructing the cell to make more viruses.

The new viruses burst out of the cell and go on to infect more cells.

The **symptoms** of disease are caused in two ways:

- bacteria and viruses produce **toxins** that interfere with the normal functioning of the body;
- viruses **damage our cells** when they burst out.

Painkillers such as paracetamol will make you feel better, but they won't kill the bacteria or virus.

● What happens when microbes get inside the body?

This is where the **immune system** is needed. There are different types of **white blood cells** whose job is to **recognise** these 'foreign' pathogens.

Pathogens are recognised because they have **antigens** on their surface, which are **not found on our cells**. In this way, our immune system can tell what is self and non-self, and attack only non-self.

Diseases caused by bacteria	Diseases caused by viruses
Food poisoning	HIV/AIDS
Cholera	Influenza
Typhoid	Cold
Leprosy	Chicken pox

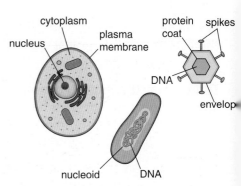

■ Simple diagrams of a human cell (left), a bacterial cell (centre) and a virus (right) (not to scale)

exam tip

★ White blood cells do not **eat** bacteria. You need a mouth and a gut to eat something. White cells **engulf** bacteria.

FOUNDATION

There are three ways in which white blood cells respond to pathogens.

- Some white blood cells make **antibodies** that attach to the pathogen so that it can be destroyed. The body makes different antibodies for different pathogens, so it makes one sort of antibody for chicken pox and another sort for measles.
- Some white blood cells produce **antitoxins** that break down the toxins (poisons) released by pathogens.
- Some white blood cells flow around pathogens (**engulfing** them), and take them into the cell where they are destroyed.

Vaccines prepare the immune system

A **vaccine stimulates** the immune system to make the right antibodies without causing the disease. Vaccines can contain **dead pathogens**, or live but **weakened (harmless) pathogens**. If a vaccinated person is then infected with a pathogen, they are **immune** to it because they already have the right antibodies.

Vaccines have saved millions of lives. However, some have **side effects**, and these could be serious. Recently a study claimed a link between the measles, mumps and rubella (**MMR**) vaccine and **autism**. Some parents decided not to give the vaccine to their children. The controversy has since died down because there is very little evidence to support the claim.

Deciding the advantages and disadvantages of vaccination means weighing up the **risks**. Ask yourself:

- What are the chances of harm if you catch the disease?
- What are the chances of being harmed by the vaccine?

With measles the chances of getting complications such as pneumonia are about one in 15, and about one in 500 die from the disease. The chances of having a side effect from the measles vaccine are about one in 1 million.

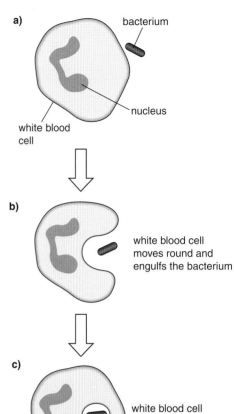

check your understanding

24 a) What is a pathogen? *(2 marks)*
  b) What are the two main types of pathogen? *(2 marks)*

25 List three ways in which your immune system responds to an infection. *(3 marks)*

26 a) What does a vaccine contain? *(1 mark)*
  b) List three diseases that babies in the UK are vaccinated against. *(3 marks)*
  c) Explain why some people don't want their baby vaccinated against a disease. *(1 mark)*

27 Doctors recommend that all children are vaccinated against measles because: *(1 mark)*

  A the MMR vaccine can cause autism.
  B the MMR vaccine doesn't hurt.
  C measles is a serious disease that can result in death.
  D measles can cause autism.

FOUNDATION

B1a Human biology

The fight against disease – then and now

Two hundred years ago, the treatment of infectious disease was not very effective. People had never heard of bacteria and viruses, and the need for cleanliness to avoid infection was not understood.

● Semmelweiss: one of the heroes in the war against disease

1840, Vienna Hospital: **Ignaz Semmelweiss** noticed that a lot of women were getting infections and dying soon after giving birth. He noticed that doctors would examine the women straight after treating other patients, or even after examining dead bodies. They did not wash their hands. Semmelweiss said that something from the dead bodies was being transferred and was causing the disease.

Semmelweiss told the doctors on his ward to **wash their hands** in **antiseptic**, and soon the death rate fell from 12% to 2%.

The antiseptic killed bacteria, although the doctors and scientists didn't know that then.

● Antibiotics and 'superbugs'

Antibiotics are medicines that kill bacteria **inside** the body. They have saved many lives. Antibiotics can't kill viruses.

Due to **over-use** of antibiotics, some strains of bacteria are now resistant to one or more antibiotics. **MRSA** stands for **methicillin-resistant *Staphylococcus aureus***. The antibiotic methicillin no longer kills that particular strain of the bacteria *Staphylococcus aureus*. The media gave the name **superbugs** to **antibiotic-resistant bacteria**.

This could be very bad news. There are more and more strains of bacteria that are resistant to more and more antibiotics. If **no** antibiotics worked, it would be like the days before antibiotics were discovered – no treatment and a high death rate.

The good news is that we are developing new antibiotics all the time. Also, doctors do not use antibiotics to treat minor infections. This **slows down** how quickly new strains of resistant bacteria develop.

> **exam tip**
>
> ★ Many students write statements such as 'the bacteria became resistant to the antibiotic', as if they suddenly decided to become resistant. Organisms can't do this. They were born 'lucky' – with the right combination of genes.

The story of antibiotic resistance is a good illustration of how evolution works

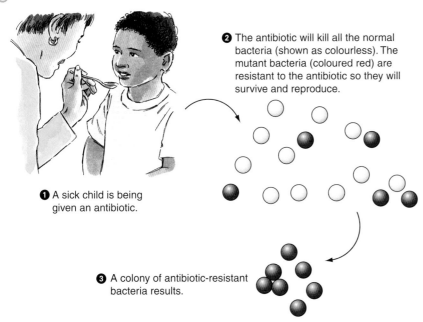

❶ A sick child is being given an antibiotic.

❷ The antibiotic will kill all the normal bacteria (shown as colourless). The mutant bacteria (coloured red) are resistant to the antibiotic so they will survive and reproduce.

❸ A colony of antibiotic-resistant bacteria results.

■ Development of a strain of bacteria resistant to an antibiotic

There are three steps to antibiotic resistance.

1 Bacteria exist in vast numbers, and they vary slightly.
2 Some bacteria are naturally resistant to penicillin. This could be due to a **mutation**, or it may be due to a gene that the bacteria have had for a long time.
3 The resistant bacteria **survived** and **reproduced**, passing their resistance genes on to the next generation. This is **natural selection**. The number of resistant bacteria increased.

Mutations and the chance of epidemics

Mutations are changes in an organism's genes. They can occur in viruses as well as bacteria. A mutated virus could make the current vaccine ineffective. This happens every year with different strains of the flu virus – a **new vaccine** has to be developed to fight the new virus strain.

If a new vaccine is not developed quickly, the disease will spread rapidly because people are not immune to it. A widespread outbreak is called an **epidemic**. An epidemic that spreads across continents or round the world is called a **pandemic**.

check your understanding

㉘ a) Why don't antibiotics work against viruses? *(1 mark)*
b) What is a 'superbug'? *(1 mark)*
c) Explain why it is important to avoid over-use of antibiotics. *(1 mark)*

㉙ In which of the following situations would you use antibiotics? *(1 mark)*

A On a person with asthma.
B On a person with the flu.
C On a person with an infected leg wound.
D On a person with cancer.

㉚ In 1957, an 'Asian flu' (influenza) virus killed 2 million people worldwide. This was a pandemic. The reason so many people died from flu is likely to be because: *(1 mark)*

A The virus was resistant to antibiotics.
B The virus mutated so that a new strain was more dangerous.
C There was no antidote to the virus.
D Not enough people took flu medicine to kill the virus.

㉛ In the nineteenth century, a scientist called Semmelweiss showed that simple hand-washing could reduce infections in his hospital. Explain how good hygiene in modern hospitals could help prevent the spread of the MRSA infection.
(2 marks)

FOUNDATION

B1b Evolution and environment

Adapt and survive

All living things need enough **warmth** and **water**. Every plant and animal also needs to get enough food, and **avoid being eaten**, so it can reproduce. Plants and animals are **adapted** to their environment.

● Polar animals are adapted so they lose less heat

- Many polar animals have a rounded shape to keep their **surface area** as **small as possible**. This minimises heat loss.
- They also have a thick layer of **blubber** that **insulates** them from the cold. This fat can also act a **food reserve**.
- Thick fur traps a layer of insulating air.
- The polar bear and arctic fox have white fur so that they are **camouflaged** – making it easier for them to hunt in snow.

■ The walrus has a thick layer of blubber so that it can swim in cold water without losing too much heat

● Desert plants are adapted to reduce water loss

Most plants need a regular supply of water, but the cactus can survive in the desert by **storing** as much water as it can, and **losing as little as possible**.

- Plants with leaves lose a lot of water from the surface of the leaves. Cacti have spines instead of leaves and a rounded shape. This **reduces the surface area water is lost from**.
- Cacti have a thick, fleshy stem that can store water, and a waxy outer layer to stop water being lost by evaporation.
- Some cacti have shallow but **far-reaching roots** to absorb water from the surface when it rains. Others have **deep roots** to reach underground water.

■ Survival in the desert is all about minimising water loss

● Desert animals are adapted to increase heat loss

Animals that live in hot, dry conditions have a problem. The best way to lose heat is by evaporation, and that means sweating. But sweating means losing water. So how do desert animals stay cool?

- Camels have **very little fat** under their skin. Instead, they have all their fat in a hump (or two) on their back. Fat all over would be an insulator and prevent the body from losing heat.
- They have thin or no hair, making heat loss easier.
- They have a long neck and legs, giving a **large surface area compared with the volume**, which increases heat loss.

■ The body shapes of desert animals are adapted to have a large surface area compared with their volume – this increases heat loss

● Avoiding predators is vital

There are many different ways to avoid being eaten.

- Some plants and animals have **weapons** and **armour** (for defence). Some plants, such as cacti, have spines or **thorns**, and animals such as hedgehogs and porcupines have spines too.
- Some animals are **camouflaged** so that predators don't notice them.
- Some animals produce **toxins** (poisons), and use **warning colours** to show this. It's no good being poisonous if a predator eats you and then spits you out – you are still dead.

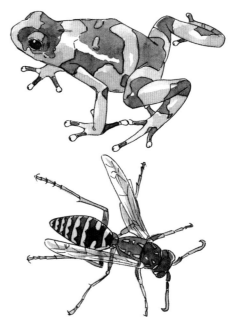

■ 'Go away, I'm poisonous' – poison arrow frog and wasp

check your understanding

1. State two ways in which a prey animal may be adapted to put off predators. *(2 marks)*

2. A sphere (ball) shape has the smallest possible surface area compared with its volume. So if an animal wants to keep warm, it helps to have a more rounded shape. Which of the following is a reason why desert animals do not have rounded shapes? *(1 mark)*

    A To run fast so they can keep warm.
    B So they can curl up in a ball to reduce heat loss.
    C To give a large surface area compared with their volume, to lose as much heat as possible.
    D To give a small surface area compared with their volume, to lose as much heat as possible.

3. Emperor penguins live in the Antarctic. Suggest why they huddle together during the coldest parts of the winter. *(2 marks)*

exam tips

★ When the examiner asks you to **state**, you only need to write a short answer – you don't need to give an explanation.

★ The relationship between surface area and heat loss often comes up, so make sure you know the basic idea.

★ You may be asked a question about a plant or animal you have not studied. Don't panic – you should be able to look at the information given, and figure out how the plant or animal is adapted to survive.

FOUNDATION

B1b Evolution and environment

Populations and competition

A **population** is a group of individuals of the same species that live in a particular environment – their **habitat**. It could be a population of elephants in a game reserve, or of oak trees in a forest.

The **size** of the population is always **limited** by a combination of:

- competition;
- predation;
- disease.

● The effect of competition

If plants or animals keep reproducing and all the offspring survive, sooner or later **resources** will be in **short supply**.

An animal might **compete** with members of its own species for:

- food;
- mates;
- territory.

And a plant might compete with other plants for:

- light;
- water;
- soil minerals.

■ They can't go on reproducing forever or food, water and space to reproduce will run out

● The effects of predators and disease

Predators such as foxes eat **prey** such as rabbits. The number of prey is affected by the number of predators, and *vice versa*. If there is a large population of rabbits in an area, there will be more food for the foxes, who will be able to raise more fox cubs. The increase in the number of foxes will then reduce the rabbit population.

Infectious disease can seriously reduce or even wipe out a population. Often, if a population is too **overcrowded**, disease will spread more rapidly. In these circumstances, only those lucky individuals whose **immune system** is good enough will survive.

● Competition between species

Different species can also compete for the **same** resources. The UK has red squirrels. Grey squirrels were introduced from the USA. In almost all areas of the UK, the population of grey squirrels has increased while the population of reds has decreased. This may be because they compete for food.

To **avoid competition**, different plants growing in the same habitat may produce flowers at **different times**. This means that there is less competition for insects to pollinate.

The effects of humans

- Humans are **predators** of some species, such as fish.
- They may also **compete** for food with some species, e.g. fishermen may compete with seals for fish.
- They may cause **pollution** that affects the supply of food for another species.
- They may **destroy habitats** by building, or **trample** plants and kill them.

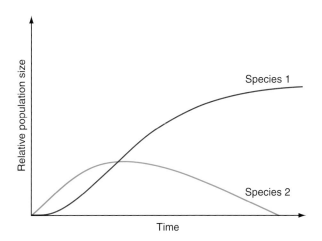

■ Populations of two plant species growing in the same area. The size of the populations can be explained if one plant grows so large that its leaves block light for the other plant. Its roots could also grow to reach most of the available water.

check your understanding

❹ The graph shows numbers of nesting pairs of peregrine falcons in the USA.
  a) Peregrine falcons have no predators. Suggest **one** reason for the decline in numbers up to 1980. *(1 mark)*
  b) Suggest **two** reasons for the recovery in the population. *(2 marks)*

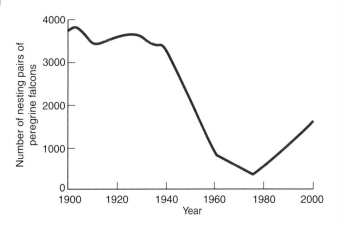

❺ Two rats are washed up on a small, deserted island. There is some food, but no other animals, so the rats begin to reproduce.
  a) Suggest why the number of rats will increase quickly at first. *(2 marks)*
  b) Suggest **two** reasons why the rate of reproduction will slow down. *(2 marks)*

❻ Bats have an adaptation called echolocation that lets them detect objects by emitting a high-pitched sound and detecting the echo. Bats eat the same prey (insects) as other birds. Which of the following is a competition advantage over other birds? *(1 mark)*

  A Bats can use the high-pitched sound to attract mates.
  B Bats can hunt at night, so competition with other predator species is reduced.
  C Bats don't bump into objects in the dark.
  D Bats can detect and avoid birds that might eat them.

FOUNDATION

B1b Evolution and environment

Variation and inheritance

It's obvious that all people are different. Even identical twins will have their differences. **Variation** is found in all other species as well.

● The ways organisms vary

■ Dogs and sheep look very different from each other because they're from different species – but sheep all vary too

Think about your year group at school. You can see variation in many different ways, such as height, weight, eye colour, shoe size, blood group and intelligence.

● There are two causes of variation

1 Your genes
- All living things are **similar** to their **parents**.
- This is because an organism's characteristics are **controlled by genes**.
- We inherit a **random mixture** of genes from **both parents**, which is why we are different from our parents, and even from our brothers and sisters.

2 Your environment
- Which means the conditions in which you live and grow up.
- This includes food you have eaten, the experiences you have had, and the diseases you have had.

So what makes you the way you are? The answer is a complicated mixture of both your genes and your environment. People often call this the **nature *versus* nurture** argument.

Some characteristics are determined just by your **genes**. Examples include eye and hair colour, blood group, and inherited diseases such as cystic fibrosis.

Other characteristics, such as height, are controlled by both genes and environment. You will inherit the genes to grow to a certain height, but you will not do so without a good diet.

FOUNDATION

Genes, chromosomes and DNA

When you first looked down a microscope at some cells, perhaps from your cheek, you will have seen the **nucleus** as just a dot. But the nucleus contains structures that hold the key to life itself.

1 The **nucleus** contains a lot of DNA. Usually it's all spread out so you can't see any detail.

2 If a cell is going to divide, it will roll its DNA up into X-shaped structures called **chromosomes**.

3 Most human cells have 23 **pairs of chromosomes**. There are two of number 1, two of number 2, etc. We can't draw them all here.

4 A chromosome is one long, tightly coiled DNA molecule. The **genes** are regions of DNA, like words on a long piece of tape. Different genes control the development of different features. There are hundreds or even thousands of genes in a single chromosome.

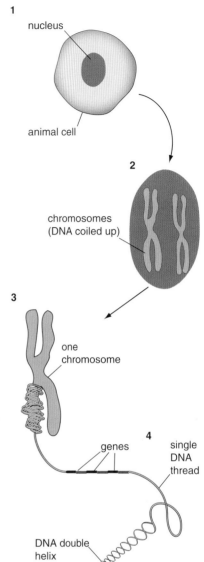

check your understanding

7 Put the following in order of size, largest first. *(2 marks)*

gene   chromosome   cell   nucleus

8 Identical twins are formed when a fertilised egg splits into two, and two babies develop separately. Which of the following is **not true** of identical twins? *(1 mark)*

A They will have the same genes.
B They will have the same colour eyes.
C They could be a boy and a girl.
D Any differences between them are probably due to their environment.

9 Which of these statements is **not** true? *(1 mark)*

A Genes are made from DNA.
B Genes are passed on in sperm and egg cells.
C Most human cells contain 23 pairs of chromosomes.
D Genes control all the illnesses you will have when you are an adult.

B1b Evolution and environment

Reproduction and cloning

Asexual reproduction produces clones

- **Asexual** means non-sexual.
- Asexual reproduction involves just **one parent**.
- There is no mixing of genetic information. All cells made asexually are **genetically identical** to the parent – they are **clones**.
- Many different species reproduce asexually, including bacteria, yeast, and many plants, such as strawberries.

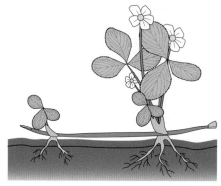

■ Asexual reproduction in a strawberry plant

Sexual reproduction gives variety

- **Sexual** reproduction involves **two parents**.
- There is a random **mixing** of **genetic information** so that each new individual is **different**.
- In sexual reproduction, the mother and the father produce **sex cells** (or **gametes**).
- The male gametes are **sperm**; the female gametes are **eggs** (**ova**).

Cloning

Identical twins result from one fertilised egg splitting into two, each of which develops into a baby. This is **natural cloning**.

Cloning plants is relatively easy.

1 Taking cuttings
Gardeners or farmers can take a cutting from a plant they want to clone. Given the right conditions, it will grow into a copy of the parent.

2 Tissue culture
A small piece of plant tissue – perhaps just a few cells – is put into a growth medium. The cells multiply, and are then separated. Each cell multiplies into a clump of cells that grows into a new plant.

Whole animals can be cloned in two ways.

Cloning animals is more difficult. Animals do not reproduce asexually, but there are ways to produce identical copies of individual farm animals, such as sheep and cattle. This can be done by **embryo transfer** or **adult cell cloning**.

a) Taking cuttings

1 Part of the stem is cut from the plant

2 The cutting is planted into compost

■ Taking cuttings

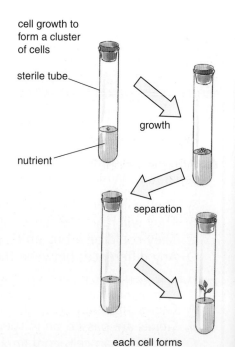

b) Tissue culture

cell growth to form a cluster of cells

sterile tube

growth

nutrient

separation

each cell forms a miniature plant

■ Tissue culture

26 FOUNDATION

This is how they made Dolly the sheep. The embryo is a clone of the adult from whom the single cell was taken.

Can we clone humans?

1 Can we? – the **science** question.
2 Should we? – the **ethics** question.

The answer to question 1 is yes – the process is just the same as cloning a sheep, but no human has been cloned yet.

The answer to question 2 depends on an individual's opinions, which may be influenced by **religious**, or **ethical beliefs**.

Cloning human tissue for treatment. Tissue culture could be used to grow cloned replacement tissue, for example skin for burns patients. Tissue could also be grown from an embryo cell that had its nucleus replaced with one from the patient. However cloning cells using embryos kills the embryo. An alternative method is **fusion cell cloning**, where a single adult cell (for example, from skin) is **fused** (joined) with cytoplasm from an egg cell. The fused cell (a clone of the original adult cell) can divide into many different types of cell.

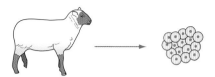

1. An embryo (clump of developing cells) is removed from a pregnant animal.

2. The embryo is split into a number of smaller clumps of cells.

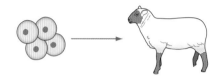

3. Each new embryo is inserted into the uterus of another host mother.

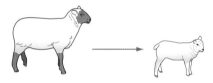

4. Some of the host mothers become pregnant and give birth to cloned offspring.

■ Embryo transfer

check your understanding

10 Describe the difference between the processes of sexual and asexual reproduction. *(1 mark)*

11 Which type of reproduction produces the most variation? Explain your answer. *(2 marks)*

12 What is a clone? *(1 mark)*

13 Give two examples of natural cloning. *(2 marks)*

14 When a gardener takes a cutting, each cutting grows into a new plant that looks identical to the original plant. The best explanation for this is: *(1 mark)*

  A the gardener used identical compost to grow the cuttings.
  B the plants have identical gametes.
  C the plants have identical genes.
  D plants of the same species do not show any variation.

exam tips

★ When the examiner asks you to **describe**, say what something looked like, or what happened, recalling facts in an accurate way. You can use diagrams as well as words.

★ When a question says **explain**, and there are two or more marks, the words **because** or **so** should appear in your answer.

FOUNDATION

B1b Evolution and environment

Genetic engineering

Using **genetic engineering**, scientists can take genes out of one species and put them into another. There are exciting possibilities, but there might be problems too.

● Genetic engineering can change a characteristic by transferring a gene

Genetic engineering involves the following process.

1. Find a gene for a useful characteristic in one organism.
2. **Cut it out** of the chromosome using an **enzyme**.
3. Cut open the DNA of the target organism.
4. **Insert** the cut-out gene into this organism's DNA.
5. Put the whole of this **modified** DNA back into a cell.
6. As the cell grows and multiplies, all the new cells have copies of the modified DNA, and so have the useful characteristic.

● Genetic engineering can produce new varieties of crops

Genes from one plant can be transferred into another species of plant. This can produce a new variety with **increased yield** (it grows bigger or more quickly), or **better nutritional value** (it may have more vitamins or protein).

Genes from bacteria can also be transferred into plants! For example, one bacterium produces a substance that kills the insects that can destroy crop plants. Scientists cut out the **insecticide gene** and transferred it into maize and cotton plants. The **genetically modified (GM)** crops then produced their own insecticide. The farmer doesn't need to spray the crop with insecticides.

● Genetic engineering can modify animals and plants as they develop

Genes can also be transferred to animal **embryos**. Genes that carry desired characteristics (such as resistance to a disease) are put into a solution. The solution is **injected** into the embryo cells. The embryo develops with the desired characteristic.

Tiny metal particles coated with genes can be fired into plant cells. The full plants grown from these plant cells have the new characteristic.

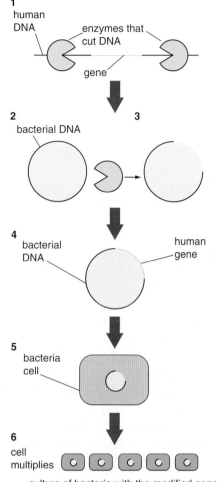

FOUNDATION

There are plenty of benefits for food production

As well as GM crops with better yield, nutrition or insect resistance, we could produce:

- crops that are **resistant** to a **plant disease**;
- crops that are **resistant** to **herbicides** (chemicals that kill plants), so the farmer can spray chemicals to kill the weeds but not the crops;
- crops that grow more quickly (by putting the growth genes from fast-growing species into slower-growing species).

But there are also problems

- Some people are concerned about the effect of **eating GM crops** on **human health**.
- Some people are worried that the genes transplanted into crops will **spread** into **wild plant populations**. This could happen if pollen transfers to other plants. There is a risk that common plants could become 'superweeds' if they get the herbicide resistance gene.
- If other (harmless) insects feed on a GM crop that kills insect pests, they will be killed.
- If farmers know they can spray weedkiller over a whole field and not harm the crop, they may use too much weedkiller and kill all the other plants, including wild flowers.

exam tip

★ Be prepared to make two lists – **advantages** and **disadvantages** of GM crops.

exam tip

If you are provided with unfamiliar information to **evaluate**, use this technique:

★ What are the pros and cons given?
★ Use the information provided, but check for people who might be biased or have unscientific opinions.
★ Use the evidence of data in graphs or tables to support your argument.

check your understanding

15. List **one** advantage to farmers of growing herbicide-resistant crops. *(1 mark)*

16. Use the following words to complete the paragraph below. *(3 marks)*

    genes   enzymes   chromosome

    Genetic engineering can modify the genes in an organism so that it has new characteristics. In genetic engineering, ____1____ from the ____2____ of one organism can be 'cut out' using ____3____ and transferred to cells of other organisms.

17. Some people are worried that growing genetically modified (GM) crops that have been modified to be resistant to herbicides will reduce the biodiversity in a field. This is because: *(1 mark)*

    A GM crops are weaker than normal crops and will die quickly.
    B GM crops could affect the health of wild plants.
    C the farmer can spray the field with weedkiller all year round, so wild flowers will die.
    D the farmer will use more pesticide, and so kill harmless insects.

FOUNDATION

B1b Evolution and environment

The fossil record

Biologists ask some very important questions:

- how did life arise?
- where do new species come from?
- why are some species extinct?

Evolution is one of the big ideas in biology.

● We're not sure how life began – nobody was there

It's important to remember that the Earth is **extremely old**. Scientists estimate that our planet is **4.6 billion** years old. That's 4 600 000 000. Scientists think life on Earth began 3.5 billion years ago.

- There are many different theories about how life began. Some are based on religion, others are based on science.
- There is a theory that life arrived on Earth as complex molecules on a **comet**. These molecules eventually became the proteins and DNA that are needed for life.
- Another theory is that these vital molecules developed on the Earth.

The first living things were like bacteria, tiny and **soft-bodied**, so they left very little **evidence** behind. As organisms evolved larger and more complex bodies, more started to be **fossilised**.

● Fossils have helped us piece it all together

Fossils are evidence that a species once existed. The fossil could be a skeleton or shell, a leaf, a footprint, or even a pile of dung.

Different layers of rock are different ages and have different fossils, so we can work out a **timeline**. We know there were fish before amphibians, for instance.

● So where did all the dinosaurs go?

There are fossils of dinosaurs, but no living dinosaurs. A species may become **extinct** if:

- a **predator** or **disease** kills them;
- the environment **changes too quickly**;
- a new species **outcompetes** them, usually for food.

FOUNDATION

Very few organisms ever turn into fossils. Most were soft-bodied, and many were eaten or just rotted away completely. Many species would have left **no trace** at all.

The very earliest fossils have been **destroyed** by geological activity. Many fossils are still **undiscovered**, or perhaps we have only **part of the skeleton**. This means there are **gaps in the fossil record**.

But the fossil record exists, and it shows that **organisms changed over time**. This is evidence for **evolution**.

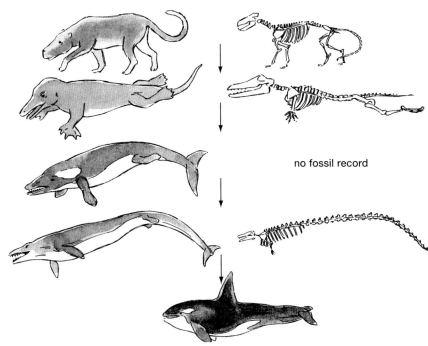

no fossil record

■ The fossil record shows that the whale may have evolved from a land-living animal

● Tracing similarities and differences

Some species, such as humans and chimpanzees, have **similar characteristics**. Sometimes the fossil record suggests that an extinct organism may be the **common ancestor** of these similar species.

An **evolutionary tree** shows how these species could be related.

check your understanding

18 The fossil record shows that some animals alive today have similar characteristics to animals that lived millions of years ago. The word for this is: *(1 mark)*

  A variation.
  B competition.
  C extinction.
  D evolution.

19 Which of the following is **not** a reason for gaps in the fossil record? *(1 mark)*

  A Most organisms never become fossilised.
  B There are many fossils still to be found.
  C Only large animals become fossilised.
  D Conditions for fossilisation occur only occasionally.

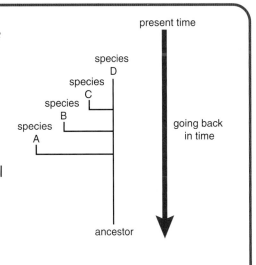

FOUNDATION

B1b Evolution and environment

How evolution happens

Evolution is usually extremely slow. But there is strong evidence from fossils that it does happen, and that if you know where to look, you can see evolution happening today.

● Evolution starts with mutations

A **mutation** is a change in an organism's **DNA**. There are three possible outcomes of a mutation:

- it may be harmful, killing the organism;
- it may have no effect at all;
- very occasionally it may be useful, for example giving better eyesight – in this case the mutation will give the individual a **survival advantage**.

● Natural selection is the mechanism of evolution

The idea of evolution had been around for centuries, but it was **Charles Darwin** who first put forward a good idea about how it could happen. He said:

- members of a population **compete** for food and mates;
- there is usually **variation** within a population due to their genes (see page 24);
- some will be born with an advantage (see the **mutation** section above) – they will be **better adapted** to their environment.
- it is more likely that this individual will **survive** and **pass on genes** to the next generation; offspring that inherit these genes will also have a survival advantage – this is **natural selection**;
- over time, this may result in an entirely new species.

● Evolution in action – rats and rat poison

In the 1950s, a new poison for rats became available – Warfarin. This was very effective and killed many rats without harming other animals or humans. But it wasn't long before warfarin-resistant rats appeared. How did that happen?

- There are a lot of rats. When Warfarin was introduced, some populations already contained rats with a chance mutation that gave them resistance to the poison.
- Warfarin was a new environmental factor that killed normal rats.
- A few resistant rats survived, reproduced and passed on the resistance gene. Rats breed very quickly. Every generation produced more resistant rats.

exam tips

★ You don't need to learn this example, but you need to be able to explain similar examples using scientific ideas. Another example is the evolution of superbugs on page 18.

★ A common mistake is to say that organisms **decide** to adapt, or mutate, after an environmental change. 'This is a nasty poison, I must become resistant to it.' This is wrong. The gene/s giving resistance **must already be there by chance**.

Darwin's theories caused great controversy

Darwin came up with his idea for evolution in his twenties, but he knew the idea would be controversial. For decades he gathered evidence to support his theory, and he did not publish it until he was about 50.

- Darwin's theory went against religious beliefs based on the Bible. This said that God made all the plants and animals living on Earth.
- Darwin did not know about DNA, genes or chromosomes, so he couldn't explain **how** characteristics could be varied or passed on. The more we find out about DNA and genes, the more the evidence supports Darwin's ideas.
- More fossils are found all the time. Gradually, more evidence accumulated to support Darwin's theory.

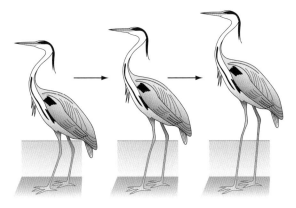

■ Lamarck's explanation for the heron's long legs

Other theories of evolution

Scientists may produce different **hypotheses** to explain similar observations. In 1809, the year Darwin was born, a Frenchman called Jean-Baptiste Lamarck put forward his theory for evolution. His idea was called the '**inheritance of acquired characteristics**'.

- Lamarck said that **useful changes**, such as longer legs, were inherited, but these changes were **acquired** during the organism's lifetime.
- We now know that the environment can change some characteristics (say by eating different food), but these changes can't be inherited.

check your understanding

20 Give one example of a characteristic acquired during an organism's lifetime that disproves Lamarck's idea. *(1 mark)*

21 Use the following words to complete the paragraph below. *(4 marks)*

        Darwin   Lamarck   inherited   acquired

Giraffes have gradually evolved long necks to reach the leaves on tall trees that no other animal can reach. ____1____'s theory explains this by saying that the giraffes' neck got longer because the giraffe's parents stretched their necks to reach the leaves. ____2____'s theory explains this by saying that a giraffe born with a long neck survived to pass on this gene to its offspring. The difference between the two theories is that Lamarck said the characteristic is ____3____, but Darwin said the characteristic is ____4____ and present from birth.

B1b Evolution and environment

How do humans affect the environment?

There are over 6 billion people on this planet. That's 6 thousand million. And the population of the world is expanding rapidly. This is because in many countries the birth rate exceeds the death rate.

● Effect on resources

1 A rising population needs more food, fuel and clean water.
2 Raw materials, including **non-renewable** energy resources, are rapidly being used up.
3 **Deforestation** is when trees are cut down for fuel or building materials, or when rainforest is replaced with land used to grow crops to produce more food.
4 Because people want a higher **standard of living**, more resources are used to make their cars, TVs or tumble dryers, or to travel to far-away places.
5 Bigger cities and more farms, landfill sites and quarries all reduce the amount of land available for animals and plants.

The world population has increased rapidly since 1800 – improvements in medicine and food production mean that more people survive to adulthood and have children

● Production of waste

Extracting metals, quarrying rocks and drilling for oil produce waste. All households produce refuse and sewage, which is dumped into land, air and sea. When human activity dumps enough stuff to upset the natural balance of ecosystems, that's **pollution.**

Pollution can kill organisms, affect **habitats** for wildlife and change the **environment**.

Air pollution:
- Carbon dioxide from **burning of fossil fuels** is thought to be causing **global warming** (see pages 36 and 54).
- **Smoke** and **soot** pollute the air we breathe.
- **Sulfur dioxide** from burning of fossil fuels can combine with water vapour in clouds to produce **acid rain**, which pollutes lakes and streams.

Land pollution:
- Pesticides and herbicides can affect wild birds, insects and flowers.
- Other **toxic chemicals** from industrial waste, such as **heavy metals**, may be washed from land into water.

Water pollution:
- **Sewage** can over-fertilise water sources, reducing supply of oxygen for aquatic creatures. (Bacteria take all the oxygen.)
- **Toxic waste** from industry, even in tiny amounts, can kill species.

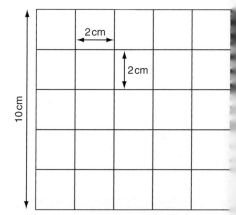
a 10cm quadrat marked on a clear sheet

1. The quadrat is placed in a few randomly chosen places.
2. The number of lichens within the quadrat, or the area covered, is counted.
3. Only about 2% of the total site needs to be sampled to give an average.

Sampling means looking at a small part of the total area

FOUNDATION

Indicators for air and water pollution

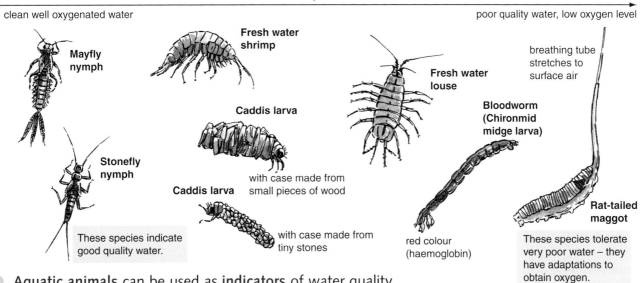

- **Aquatic animals** can be used as **indicators** of water quality.
- **Lichens** are very sensitive to pollutants in the atmosphere. Where they do grow, they indicate clean air.

■ Some organisms thrive when dissolved oxygen levels are low – their presence indicates polluted water

Suppose you were **designing an investigation** to **compare** the numbers of indicator species in two different locations.

- The more **samples** taken, the more **reliable** the estimate will be.
- Samples must be taken from **random locations** – this means the sample is likely to be **representative**, and you are not introducing **bias**.
- Expect a small variation in sample results due to the random sampling, but an **anomalous** result may not be representative.
- A **consistently large difference** in the measurement by different groups, at the same site, could be due to a **systematic error**. For example, did the two groups use different-sized buckets for sampling?

check your understanding

22. Name **two** non-renewable raw materials. (2 marks)

23. A fieldwork survey used quadrats to estimate the percentage cover of lichens growing on tree trunks in different areas of the school grounds. A student suggested sampling the trees nearest to the road, and comparing them with those furthest from the road.
    a) Explain what information you can get from the growth of lichen. (2 marks)
    b) Suggest what the student's hypothesis (idea) was. (1 mark)
    c) Explain why it was important to sample more than one tree in each area. (2 marks)

24. Use the following words to complete the paragraph below. (4 marks)

    deforestation   acid rain   global warming   sewage

    Burning wood releases sulfur dioxide, which causes ____1____ and carbon dioxide, which causes ____2____. The population is increasing, so uses more fuel and produces more ____3____ as well as causing ____4____.

FOUNDATION

B1b Evolution and environment

Global warming

The Earth has been getting warmer in the last 200 years (**global warming**). Scientists think this is due to an increased **greenhouse effect**: as in a greenhouse, the atmosphere is allowing sunlight **energy in** but not letting all of it escape, so it gets hotter.

● What causes the greenhouse effect?

- The temperatures on Earth result from a **balance** in **energy received** and **energy lost**.
- The energy we get from the Sun is not changing, but we are **losing less**.
- This is because we are **changing** our **atmosphere**. There's more **carbon dioxide** and **methane** (**greenhouse gases**). These gases prevent heat from escaping back out into space.

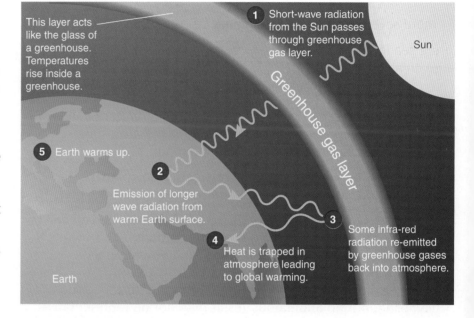

Where does the carbon dioxide come from?

Burning fossil fuels (coal, gas, oil) produces millions of tonnes of carbon dioxide every year. More industry means more **combustion** and more carbon dioxide.

However, only **photosynthesis** takes carbon dioxide **out** of the atmosphere.

There's a lot of carbon locked up in wood, as well as fossil fuels. Cutting down forests – **deforestation** – and burning them means:

- carbon dioxide is released in burning;
- there are fewer trees to remove carbon dioxide from the atmosphere when they photosynthesise.

Methane gas (CH_4) is another greenhouse gas. Methane production is increasing due to:

- cows belching and farting (**flatulence**) – as more cattle are being bred to produce more food, more methane is produced;
- rice paddy fields produce methane – more rice is being grown as the world's population grows.

FOUNDATION

The effects of global warming are complex

The Earth's climate is a **complex system**. That means it's impossible to predict exactly what will happen. Here are some possibilities:

- higher temperatures make ice melt – **glaciers** and **ice caps** on Greenland and Antarctica will **melt** and drain into the sea, making the **sea level rise**;
- higher temperatures make sea water **expand**, making the sea level rise;
- the extra fresh water may **disrupt ocean currents** – this could be a disaster for Britain, currently kept warmer by Atlantic currents;
- more severe weather – there could be more hurricanes.

Rising sea levels will flood low-lying areas such as East Anglia and large areas of the Netherlands. Millions of people may be homeless.

All over the world, ecosystems depend on the climate. Changes could disrupt plant growth and the pattern of agriculture could change.

■ The Maldives are islands in the Indian Ocean. The islands are mostly less than 1m above sea level. The most gloomy predictions say that the Maldives could be totally flooded within the next few decades.

How can we be sure?

All the evidence shows that global temperatures are rising and that ice caps are melting. The evidence also shows that all this is due to human activity.

- There is a **correlation** between increased greenhouse gases and global warming.
- A correlation might be just a **coincidence**, but the rising temperatures agree with **climate prediction models** carried out by **different scientists** from all over the world, making the conclusion more **plausible**.

The computer models also show that the effect of a rise of even just a few degrees will be damaging. We may not feel the difference in our lifetime, but it is clear that changes are happening, so we must do something for future generations.

check your understanding

25. List **two** factors contributing to increased concentrations of methane in the atmosphere. *(2 marks)*
26. State what effect deforestation has on concentrations of carbon dioxide in the atmosphere, and explain your answer. *(2 marks)*
27. The greenhouse effect is caused by: *(1 mark)*

   A a layer of gas around the Earth that stops the Sun's radiation from reaching the Earth.
   B a layer of gas around the Earth that increases the amount of the Sun's radiation reaching the Earth.
   C a layer of gas around the Earth that traps energy radiated by the Sun before it reaches the Earth's surface.
   D a layer of gas around the Earth that traps energy radiated by the Sun after it is re-radiated from the Earth's surface.

FOUNDATION

B1b Evolution and environment

Sustainable development

A present, we are doing many things that are not **sustainable**, which means that it can't go on for ever:

- using fossil fuels, which are **non-renewable** and will eventually **run out**;
- using up raw material resources, such as metal ores;
- polluting ecosystems, such as rainforests and coral reefs, which will not grow back.

Sustainable development meets the needs of today's generation without damaging the planet for future generations.

● Sustainable development needs planning and co-operation

Planning is needed at global, regional and local levels.

Governments need to **co-operate** and agree on **policies**. There has been some success in **international agreements**. For example:

- many countries have agreed targets to reduce carbon dioxide emissions (the **Kyoto Agreement**);
- many countries have signed up to **quotas** on **cod fishing**, which will **conserve** fish stocks.

Similar agreements could help **endangered species** that are essential for use as food or medicines.

One problem is that we don't yet know which of the species that have been lost due to **deforestation** in the Amazon may have been of future use, or what **undiscovered species** could have potential benefits.

At a regional level:

- better **public transport** would reduce the number of car journeys;
- planning agreements for land use could prevent **loss of wildlife habitats**.

At the local level (your home, your school, your town):

- **recycling paper** means that fewer forests are cut down; **recycling glass and metals** means that fewer quarries are dug;
- using energy **more efficiently conserves** supplies of fossil fuels and **reduces pollution**;
- using **renewable** energy resources also **conserves** supplies of fossil fuels and **reduces pollution**.

check your understanding

28. Give two reasons why recycling should be part of a local council's sustainable development plan. *(2 marks)*

29. Which of the following is **not** a reason why national and international agreements on stopping the destruction of rainforests are important? *(1 mark)*

   A They would conserve supplies of fossil fuels.
   B They would reduce carbon dioxide emissions.
   C The rainforests may be important for discovering new medicines.
   D They would maintain biodiversity.

FOUNDATION

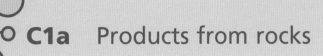

C1a Products from rocks

Atoms, elements and the Periodic Table

● All about atoms

- All substances are made of **atoms**.
- Atoms have a tiny central **nucleus**, around which there are **electrons**.
- Most of an atom is empty space: the electrons are a long way from the nucleus.

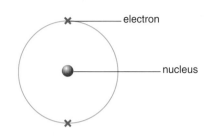

■ An atom: a nucleus surrounded by electrons

● All about elements

- A substance that is made of only one sort of atom is called an **element**.
- A gold bar contains just gold atoms, and nothing else.
- A **chemical symbol** represents one atom of an element.
- Each element has a different symbol, which could be one or two letters. O represents an atom of oxygen; Na represents an atom of sodium.
- When elements react, their atoms join with other atoms to form **compounds**.
- For example: sodium and oxygen react to form sodium oxide.

exam tips

★ Learn the definition of an element – it could be worth *2 marks* (*1 mark* for atom, *1 mark* for element).

★ There are only about 100 different **elements**.

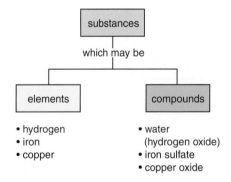

● All about the Periodic Table

- These elements are shown in the **Periodic Table**.
- The columns (called **groups**) contain elements with **similar properties** (look at Group 1).

FOUNDATION

1 Alkali metals	2 Alkaline-earth metals										3	4	5	6	7 Halogens	0 Noble gases
H hydrogen 1																He helium 2
Li lithium 3	Be beryllium 4										B boron 5	C carbon 6	N nitrogen 7	O oxygen 8	F fluorine 9	Ne neon 10
Na sodium 11	Mg magnesium 12			transition elements							Al aluminium 13	Si silicon 14	P phosphorus 15	S sulfur 16	Cl chlorine 17	Ar argon 18
K potassium 19	Ca calcium 20	Sc 21	Ti 22	V 23	Cr chromium 24	Mn manganese 25	Fe iron 26	Co cobalt 27	Ni nickel 28	Cu copper 29	Zn zinc 30	Ga 31	Ge 32	As 33	Se 34	Br bromine 35	Kr krypton 36
Rb 37	Sr 38	Y 39	Zr 40	Nb 41	Mo 42	Tc 43	Ru 44	Rh 45	Pd 46	Ag silver 47	Cd 48	In 49	Sn tin 50	Sb 51	Te 52	I iodine 53	Xe 54
Cs 55	Ba 56	La 57	Hf 72	Ta 73	W 74	Re 75	Os 76	Ir 77	Pt platinum 78	Au gold 79	Hg mercury 80	Tl 81	Pb lead 82	Bi 83	Po 84	At 85	Rn 86
Fr 87	Ra 88	Ac 89	Rf 104	Db 105	Sg 106	Bh 107	Hs 108	Mt 109								

Key: H hydrogen 1 ← symbol ← name ← atomic number

■ The Periodic Table

So all you have to know are these four points:

1 Everything is made of **atoms**, which all have a **nucleus** surrounded by **electrons**.
2 Something made of just one type of atom is an **element** (like gold, for example), and each of the 100 elements has been given a **chemical symbol**.
3 Elements combine to form **compounds**.
4 The elements are organised into the **Periodic Table** – elements with **similar properties** are found in the same **group** or column.

check your understanding

1. Draw and label a simple diagram of an atom. *(3 marks)*
2. If some moon rock was found to be made of one sort of atom, would it be an element or a compound? *(1 mark)*
3. Which sub-structure of the Periodic Table shows elements that have similar properties? *(1 mark)*

   A families B periods C groups D gatherings

4. Which of these chemical symbols is **not** used in the Periodic Table? *(1 mark)*

   A Al B Cl C Fl D Tl

FOUNDATION

C1a   Products from rocks

Reactions, formulae and balanced equations

● Compounds

- Atoms join with other atoms to make **chemical bonds** in compounds.
- Electrons can be **transferred** from one atom to another, or **shared** between the atoms.
- **Electron sharing** between atoms produces molecules.

In a molecule of oxygen (O_2), there are two atoms of the same type, so this is an element.

In a molecule of water (H_2O), there is more than one type of atom, so this is a compound.

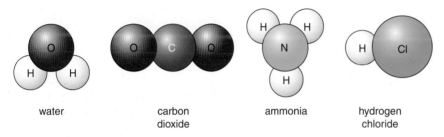

■ Atoms join together to make compounds

● Formulae

We use atomic names and **chemical symbols** to show what is happening in chemical reactions – like this:

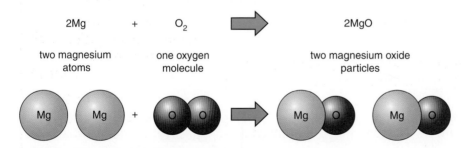

The **formula** of a compound shows the number and type of atoms that are joined together.

- The formula for oxygen is O_2 – two atoms of the element for which the symbol is O.
- The formula for magnesium oxide is MgO – one atom of magnesium (Mg) and one atom of oxygen (O).

42   FOUNDATION

Formulae in equations

The **word equation** for the reaction to form magnesium oxide is:

magnesium + oxygen → magnesium oxide

If we just write the **formulae** for the reactants and products:

$Mg + O_2 \rightarrow MgO$

the equation is not **balanced**. There are two oxygen atoms on the left-hand side, but only one oxygen atom on the right-hand side.

So MgO has to be doubled to give:

$Mg + O_2 \rightarrow 2MgO$

But now we have two Mg atoms on the right-hand side, and only one on the other side. We correct this by putting 2Mg on the left:

$2Mg + O_2 \rightarrow 2MgO$

This is the **balanced chemical equation**.

A balanced chemical equation has the same number of the same atoms on each side – they are just arranged differently.

Balancing equations – never change the formula of a compound!

- The formula of a compound tells you exactly the number and type of atoms in it.
- You **can't lose or make atoms** during a chemical reaction.
- So the **mass of what's produced** has to be the same as the **mass of the starting materials**.
- A **balanced equation** shows the number and type of atoms involved.

Remember: the formula of magnesium oxide is MgO, showing one atom of magnesium has joined with one atom of oxygen.

In the equation:

$2Mg + O_2 \rightarrow 2MgO$

2MgO tells us that two units of MgO are made.

2Mg tells us that two **atoms** of magnesium are used.

O_2 tells us that one **molecule** of oxygen is used.

Remember to **count the atoms on both sides**.

---

check your understanding

5 Copy the following sentences and fill in the missing words, selecting from the words listed below: *(3 marks)*

> bonds   compound
> electrons   molecules
> nucleus

One or more atoms can be held together by chemical ____1____. The ____2____ in the atom can be shared or transferred. When atoms of two or more elements are joined, a ____3____ is formed.

6 In the equation for the reaction of calcium with oxygen to form calcium oxide:

$2Ca + O_2 \rightarrow 2CaO$

what are the correct numbers of atoms on each side of the equation?
(1 mark)

A 2 on the left and 2 on the right
B 2 on the left and 1 on the right
C 3 on the left and 2 on the right
D 4 on the left and 4 on the right

FOUNDATION

C1a Products from rocks

Products from limestone

Limestone, mainly calcium carbonate ($CaCO_3$), is **quarried** and can be used as a **building material**.

How many uses? Just six to remember:

- **building blocks** – in buildings;
- **slaked lime** – to combat acidity;
- **mortar** – to stick bricks together;
- **cement** – to make mortar;
- **concrete** – for foundations;
- **glass** – in windows, drinking glasses.

■ A limestone quarry

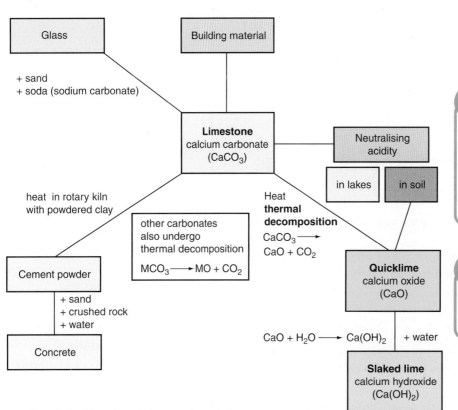

■ Uses of limestone

> **exam tip**
> ★ Limestone is used **as** building blocks, but it is used to **make** the other materials. Take care what you say.

> **exam tip**
> ★ Learn the equations in this summary diagram.

● Making limestone useful

To do anything with limestone, other than use it as the rock, requires chemistry to modify its properties.

- **Building** blocks – quarried and cut to shape.
- **Cement** – heat limestone with clay.
- **Mortar** – mix cement with sand and water.
- **Concrete** – mix mortar with small stones (aggregate).
- **Glass** – heat a mixture of limestone, sand and sodium carbonate, which makes soda glass, one type of glass.

FOUNDATION

Modifying these materials

Different sorts of glass, such as coloured and toughened glass, are made by adding different chemicals to the molten glass mixture. Even a **smart material**, such as the glass in the type of sunglasses that react to light, is made like this.

Concrete is very strong when it is squeezed, but quite weak if something is trying to snap it. **Reinforcing** it with steel rods produces a material that is strong under all types of strain.

Thermal decomposition is heating something to break it into simpler materials.

The diagram on page 44 (Uses of limestone) shows that heat will turn calcium carbonate into calcium oxide (quicklime) and carbon dioxide.

The same thing happens for other metal carbonates:
$MCO_3 \rightarrow MO + CO_2$

In this equation, M stands for a metal. Metal carbonates decompose on heating – to form carbon dioxide and a metal oxide.

Advantages and disadvantages of using these products

Here are some factors to consider:

- Glass is easily broken.
- Glass does not scratch easily and is waterproof.
- Glass buildings let in a lot of light, but can overheat in summer.
- Limestone buildings are good to look at as limestone can be carved and polishes easily.
- Limestone is fairly cheap and so are its products.
- Concrete can be poured into moulds to make different shapes.
- Limestone, concrete, and bricks and mortar are stronger than timber buildings.

check your understanding

7 Write the word equation for the thermal decomposition of copper carbonate. *(2 marks)*

8 Calcium hydroxide (slaked lime) has the formula $Ca(OH)_2$. How many atoms are there in a molecule of calcium hydroxide? *(1 mark)*

   A 3   B 4   C 5   D 6

9 Which of the following reasons may be given for using limestone, as opposed to granite or concrete, as an external building material? *(1 mark)*

   A It is difficult to carve.
   B Liquid limestone can be poured into moulds.
   C It polishes easily.
   D It can be modified by the addition of small amounts of other chemicals.

C1a Products from rocks

Extracting metals

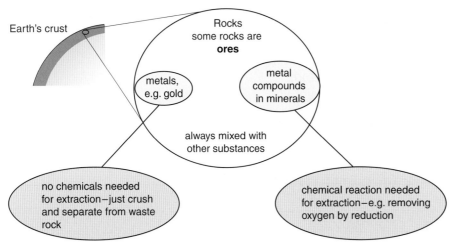

■ Metals and metal compounds in the Earth's crust

● The reactivity series

- Unreactive metals, such as gold, are found in the Earth as the metal itself.
- Reactive metals are found as minerals in **ores**. Minerals are compounds, such as oxides or carbonates, so a chemical reaction is needed to extract the metal.
- Metals that are **less reactive** than carbon can be extracted from their oxides by **reduction** with carbon.
- Iron oxide is reduced with **coke** in the blast furnace to make iron.
- Aluminium and titanium cannot be extracted from their oxides by reduction with carbon (they are **too reactive**).

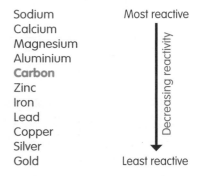

■ The reactivity series – note aluminium is above carbon, but iron is below carbon

● Purifying copper using electrolysis

- Copper could be extracted from copper oxide using carbon.
- However, the copper produced this way is impure. Copper has to be **very pure** to be a good conductor for use in electrical wiring.
- **Electrolysis** is used to extract copper from solutions containing copper compounds, but it uses large amounts of electricity, which is expensive.

● Aluminium – a valuable metal

- Aluminium is also extracted by electrolysis.
- Electrolysis is expensive because a lot of energy is needed.
- The processes of mining and concentrating the ore are also expensive.

46 FOUNDATION

- However, aluminium is so valuable that, even if the ore contains only a low percentage of metal, it is still economic to extract the metal.
- If the price of the metal goes down, it may not be economical to extract it from a **low-grade** ore.

Copper – the useful metal

- Copper has many uses, but the supply of copper-rich ores is limited.
- We need to find new ways of extracting copper from low-grade ores.
- Such methods include using bacteria, fungi and plants.
- Extracting copper from waste tips at old mines would be 'greener' than starting a new mine.

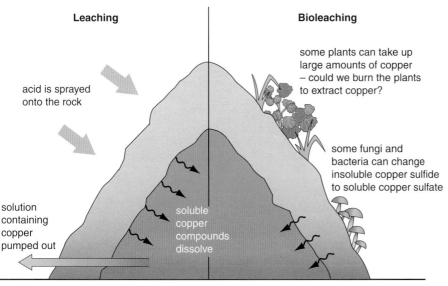

■ New methods of extracting copper from copper sulfide ores are slow, but use very little energy and do not cause sulfur dioxide emissions

check your understanding

10 Which metals are found as themselves in the Earth – the least reactive or the most reactive? *(1 mark)*

11 Which type of reaction is used in the blast furnace to extract iron? *(1 mark)*

　A reduction　　　　B thermal decomposition
　C extraction　　　　D quickliming

12 Which of the following metals cannot be extracted by reduction with carbon? *(1 mark)*

　A aluminium　　　　B copper
　C iron　　　　　　　D lead

exam tip

★ Work out which ones are the **wrong** answers and cross them out as you go. Then write down the correct answer.

C1a Products from rocks

Quarrying, mining and recycling

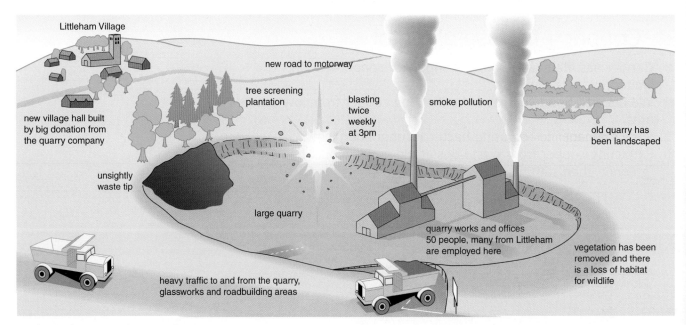

■ A typical quarry and surrounding area

Look at the picture. Imagine what you would think of the quarry if you lived in the village.

Your **opinions** for or against quarrying may be:

- the quarry provides employment for your family and friends;
- the hole is ugly;
- the quarry provides necessary rock for producing building materials;
- there has been a loss of wildlife habitat;
- the lorries are dirty;
- the blasting is noisy;
- the works produce a lot of smoke pollution.

Do you have **evidence** to support all your opinions?

Taking public opinion into account, the quarry company might try to reduce the **environmental, economic and social impact** of the new quarry by:

- washing the lorries before they leave the site;
- blasting only twice a week;
- cleaning the smoke before it goes up the chimney;
- building new roads to link up with motorways;
- building a social centre/village hall;
- landscaping the site afterwards.

exam tip

★ Be prepared to make two lists – **advantages** and **disadvantages** of a new quarry or mine.

Mining and recycling

- Like quarrying, mining also means digging out a large amount of rock, and produces large amounts of waste.
- We should **recycle** metals because extracting them spoils **the environment**, uses **limited resources** and uses **large amounts of energy**.
- Recycling uses energy too, but generally uses less than extracting metals from ores using heat or electricity.

■ A car contains copper wiring and a radiator made from copper and brass. The total mass of copper is between 15 and 25 kg. This can all be recycled.

Why recycle?

Price – it's cheaper to recycle than dig out more ore.

Limited resources – it makes sense to make what we've got last longer.

Energy efficiency – recycling uses 85% less energy.

Landfill costs – problems of rubbish disposal are avoided by recycling.

The environment – no mining means fewer spoil heaps, and there are virtually no emissions of harmful gases (some methods of processing copper ore release sulfur dioxide).

Recycling difficulties

- Separating metals is very tricky.
- Iron and steel, however, are easily separated from other metals – with a magnet.

check your understanding

13 List three benefits of quarrying limestone. *(3 marks)*

14 List three drawbacks of quarrying limestone. *(3 marks)*

15 From this list, which is **not** a reason why we should recycle metals? *(1 mark)*

    A It's cheaper to recycle.
    B It conserves (keeps) remaining ore.
    C It uses more energy.
    D It saves valuable landfill space.

exam tip

★ If the question asks you to **list**, you should give concise answers in a list, one after the other.

FOUNDATION

C1a Products from rocks

Using metals and alloys

Iron and steel

- Iron from the blast furnace (sometimes used as **cast iron**) contains about 96% iron.
- The impurities make it **brittle** and so cast iron isn't very useful.
- Pure iron is 100% iron, but is too soft to be very useful.
- Most iron is converted into **steels**.
- Steels are **alloys** – **mixtures** of iron, carbon and other metals.

Properties of alloys

In **alloys**, the different-sized atoms disrupt the pattern of layers. This makes it more difficult for the atoms to **slide** over each other, so alloys are **harder** than pure metals.

Alloys can be designed to have particular properties by varying the amounts of the other metals that are added.

The amount of carbon has the greatest effect on the properties of steel:

- **low-carbon steels** are easy to make into different shapes;
- **high-carbon steels** are hard, but brittle;
- **stainless steels** don't rust very easily.

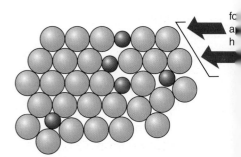

■ Pure iron is soft because layers of atoms can slide over each other – slip cannot occur so easily in an alloy, because the regular structure of atoms has been distorted

The transition metals

- The **transition metals** are typical metals.
- They are good conductors of heat and electricity, and can be bent or hammered into shape (**malleable**).
- They are useful as structural (building) materials.
- Copper is useful for electrical wiring and plumbing because:
  - it conducts electricity well;
  - it is easily drawn into wires or tubes (**ductile**).

exam tip

★ If the question asks you to **'give a reason'** you need to apply scientific knowledge, but you can use short sentences or bullet points like these.

■ Transition metals are the central block of the Periodic Table

FOUNDATION

Aluminium and titanium

- **Low density** makes aluminium and titanium useful. Lightweight jet planes made from aluminium and titanium can travel faster.
- Their **resistance to corrosion** is also useful. Aluminium window frames don't corrode, so they don't need painting to protect them from air and water.

■ Aluminium and titanium are used to make aircraft parts because they resist corrosion and have low densities

Everyday metals

Many metals in everyday use are alloys. The pure metal would be too soft, so small amounts of similar metals are added to make them harder:

- stainless steel is often used to make knives, forks and spoons.
- coins are alloys of copper and nickel.
- brass is an alloy of copper and zinc.
- pure gold rings would be far too soft to keep their shape, so copper is added to increase their strength.

Some new spectacle frames and dental braces are made of a **shape-memory alloy** – **smart alloys** can return to their original shape after being twisted.

exam tips

★ Make sure you say 'lightweight', not just 'light'. After all, a tonne of steel weighs the same as a tonne of aluminium. Even better, say 'low density'.

★ In the exam, you may have to draw conclusions about why you would use what metal. You need to consider the metals' properties.

Comparing different metals

Iron from the blast furnace is strong in compression, but brittle. It would be good for manhole covers, but dangerous for building bridges and girders.

High-carbon steel is also brittle, so it is not suitable for use as steel girders. However, it is good to use for drill bits, which need to be hard and stay sharp. Low-carbon steel would be too soft.

Copper and aluminium are soft, so their harder alloys have many more uses than the pure metals.

check your understanding

16 How do the impurities in iron from the blast affect the properties of iron? *(1 mark)*

17 Which of the following statements is the main advantage of making coins from an alloy of copper instead of just copper? *(1 mark)*

   A The alloy is shinier.   B The alloy is harder.
   C The alloy does not conduct heat.   D The alloy does not conduct electricity.

18 What is the really useful property of shape-memory alloys? *(1 mark)*

   A They remember their original colour.   B They remember their original density.
   C They remember their original strength. D They remember their original shape.

C1a Products from rocks

Crude oil and fuels

- Crude oil is a **mixture** of many compounds, mainly **hydrocarbons**.
- **Hydrocarbons** are molecules made up of hydrogen and carbon atoms **only** – such as methane, CH_4.
- A **mixture** consists of two or more elements or compounds **not chemically combined** together. The chemical properties of each substance in the mixture are unchanged.
- It is possible to **separate** the substances in a mixture by physical methods, including **distillation**.

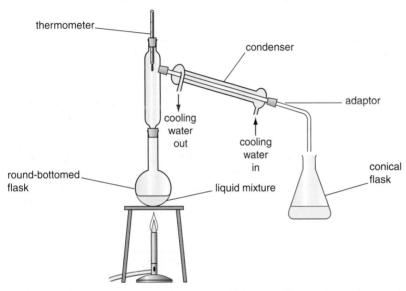

■ Distillation – there is just one temperature at which the boiling liquid condenses

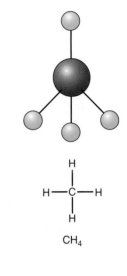

● Alkanes

- Most of the compounds in crude oil are **saturated hydrocarbons** called **alkanes**.
- **Saturated** means that you can't fit any more hydrogens into the molecule.
- **Alkane** molecules can be represented in the forms shown in the diagram. Can you see that no more hydrogens will fit in?

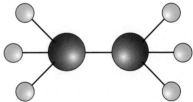

● Fractional distillation

- In **fractional distillation**, many liquids of different **boiling points** can be separated into **fractions** (see table opposite).
- The hydrocarbons in crude oil can be **separated** into fractions.
- Each fraction is made up of molecules with a similar number of carbon atoms.
- They are separated by heating up and evaporating the oil and allowing it to **condense** (turn back to liquid) at different temperatures.
- The different fractions are easily separated because of their different boiling points.

exam tips

★ You don't need to learn the names of all the alkanes, but try to recognise that it's an alkane like CH_4 from the formula C_nH_{2n+2}.

★ Make sure you know the difference between **distillation** and **fractional distillation**.

FOUNDATION

Fractions from an oil refinery					
Fraction	Number of carbon atoms in the molecule	Description and viscosity	Flammability	Boiling point in °C	Uses
Refinery gas	1–4	Colourless gases	Explodes if mixed with air and lit	Less than 40	Used as a fuel in the refinery. Bottled and sold as LPG
Naphtha	5–10	Yellowish liquid flows very easily	Evaporates easily, vapour mixed with air is explosive	25–175	Petrol
Kerosene	10–14	Yellowish liquid flows like water	Will burn when heated	150–260	Aircraft fuel
Light gas oil	14–20	Yellow liquid thicker than water	Needs soaking onto a wick or other material to burn	235–360	Diesel fuel
Heavy gas oil	20–50	Yellow brown liquid	Just burns when soaked onto a wick – very smoky	330–380	Fuel for boilers and heaters
Fuel oil	60–80	Thick brown sticky liquid	Needs to be hot and soaked onto a wick before it will burn	above 490	Fuel oil for power stations and ships

Fuels

Some properties of **hydrocarbons**, such as **viscosity**, **flammability** and **boiling point**, depend on how big the molecules are. These properties help to decide how hydrocarbons are used as **fuels**.

You can find the following from the table:

- each fraction has a **range** of molecule sizes and a **range** of **boiling points**;
- larger molecules have higher boiling points.

exam tip

★ You don't need to learn the names in the table, but you need to be able to compare data and information about the properties of fuels.

check your understanding

19 Which of the following is **not** a mixture? *(1 mark)*

　A sea water　　B air　　C crude oil　　D salt

20 Crude oil can be separated into more useful: *(1 mark)*

　A parts.　　B sections.　　C decimals.　　D fractions.

21 Which of these is **not** a hydrocarbon? *(1 mark)*

　A CH_4　　B C_2H_5OH　　C CH_3CH_3　　D C_2H_6

22 Use the table to describe how the colour of the fractions changes as the molecules get bigger. *(1 mark)*

FOUNDATION

C1a Products from rocks

Burning fuels

● Products of combustion

Burning fuels (**combustion**) gives us energy, but also produces other substances.

- Most fuels are hydrocarbons and so contain carbon and hydrogen. Some fuels also contain some sulfur.
- So gases released into the atmosphere when a fuel burns may include **carbon dioxide**, **water** (vapour), **carbon monoxide** and **sulfur dioxide**.

This apparatus shows us the products of burning a hydrocarbon. The water vapour **condenses** in the cold U-tube and the carbon dioxide turns the **limewater milky**.

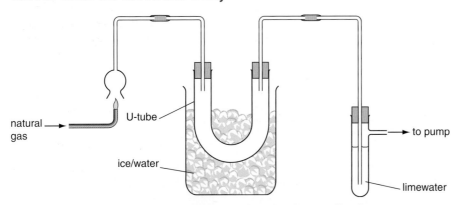

■ Hydrocarbons burn to produce water and carbon dioxide as well as heat energy

Methane, the gas used in most domestic central heating boilers, burns like this:

$$CH_4 + 2O_2 \rightarrow CO_2 + 2H_2O$$
$$\text{methane} + \text{oxygen} \rightarrow \text{carbon dioxide} + \text{water}$$

When an **alkane** burns in **plenty** of air (**oxygen**), **only** carbon dioxide and water are produced.

● Harmful emissions from burning fuels

- **Particles** (such as **soot** or **unburnt hydrocarbons**) may be released, especially if there is not enough oxygen for complete combustion (as in a car engine).
- Particles cause **global dimming** – sunlight cannot get through the atmosphere as easily.
- Sulfur impurities burn to produce sulfur dioxide, which causes **acid rain** – dilute sulfuric acid.
- Carbon dioxide emissions cause **global warming** – too much greenhouse effect.
- Combustion can also produce **carbon monoxide**, which is poisonous.

> **exam tip**
> ★ If you're asked what is produced when a certain fuel burns (the fuel will be an alkane) the answer is 'carbon dioxide plus water'.

Impact on the environment

What happens when we burn fossil fuels? The diagram says it all!

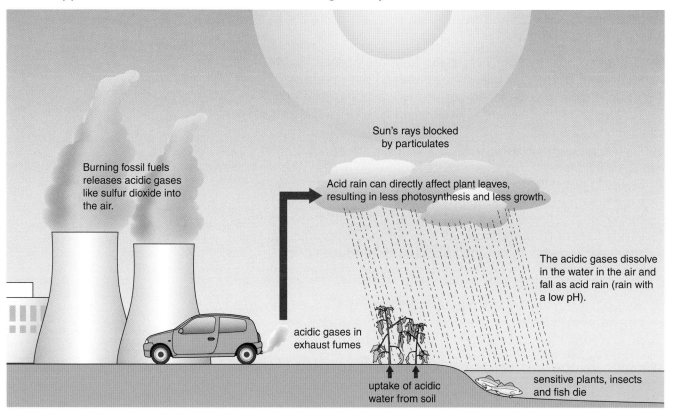

■ The environmental impacts of burning fossil fuels

Most scientists agree that the **data** show that global warming is due to increased carbon dioxide emissions. **Opinions** about other social, economic and environmental impacts of the uses of fuels could be influenced by **bias**.

check your understanding

23 Explain why one product of combustion could cause global temperatures to decrease. *(3 marks)*

24 The products of burning a pure alkane are: *(1 mark)*

   A carbon dioxide and water.
   B carbon monoxide and water.
   C sulfur dioxide and hydrogen.
   D sulfur dioxide and water.

25 If not enough oxygen is present, combustion may not be complete. Which of these gases forms only when combustion is not complete? *(1 mark)*

   A water vapour   B sulfur dioxide   C carbon dioxide   D carbon monoxide

exam tip

★ If the question says **explain**, you should not just give a list of reasons – you need to use some theory to give a reason why something is happening.

FOUNDATION

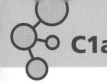

C1a Products from rocks

Cleaner fuels

Sulfur in fuel is a problem. It causes acid rain. But there are two ways of preventing that.

- Sulfur can be taken out **before** the fuels are burned. So we have **low-sulfur petrol** and **low-sulfur diesel** to put in our cars.
- Sulfur dioxide can be taken out of the waste gases at power stations **after** combustion. This is called **flue gas desulfurisation**.

● Alternative fuels

- Fuels do not have to come from crude oil. Plants, rubbish and even cow dung are used to produce **biofuels**. **Ethanol** and **hydrogen** are two biofuels used as clean alternatives to petrol.
- Plants **absorb** carbon dioxide as they grow. Crops, like sugar cane, are **fermented** and then **distilled** to produce **ethanol**. When this burns, the original carbon dioxide is released and water is also produced.
- **Hydrogen** is made by splitting up water and can be used in an engine or in **fuel cells** to produce electricity. Water is also produced whenever hydrogen is used as a fuel.

What are the social, economic and environmental impacts of using these alternative fuels?

Ethanol and hydrogen have advantages over hydrocarbons as fuels.

- They do not deplete **non-renewable** crude oil reserves.
- They are 'clean' – they produce no toxic fumes, and using hydrogen to generate electricity does not produce carbon dioxide.

The drawback of ethanol is that it takes a lot of sugar cane to make the fuel – the fields could be used to produce food. Compared with Brazil, where a lot of ethanol is produced for fuel, Europe has less land available and a less sunny climate.

There is a problem **storing** enough hydrogen, as it is a gas and takes up a large volume. It needs to be pressurised or liquefied, which takes a lot of energy.

> **exam tip**
> ★ If you are asked for economic and environmental effects of using fuels, you must remember to give a **balanced** argument and deal with **both** aspects – forget one and you will lose half your marks.

■ Do you think these people are right? Are alternative / cleaner fuels the whole answer?

● Energy outputs

The trouble with some cleaner fuels is that they do not give out as much energy as dirty old diesel. Look at the table and decide which one you would like to put into your 'petrol tank'. Which one will take you the furthest?

	Diesel	Biodiesel	Ethanol	Hydrogen (liquid)
Energy content per gallon (British thermal units)	130 000	120 000	80 000	30 500

check your understanding

㉖ Flue gas desulfurisation removes which one of the following from industrial waste gases? *(1 mark)*

    A sulfur gas
    B hydrogen sulfide
    C sulfur particles
    D sulfur dioxide

㉗ List one advantage and one disadvantage of using ethanol as a fuel. *(2 marks)*

㉘ Write the word equation for the combustion of ethanol. *(2 marks)*

C1b Oils, Earth and atmosphere

Cracking crude oil

- The hydrocarbons from **fractional distillation** of crude oil can be broken down (**cracked**) to produce smaller molecules.
- These molecules are very useful.
- **Cracking** involves heating the **hydrocarbons** to turn them into gas, and passing the gas over a hot **catalyst**.
- A **catalyst** speeds up a reaction but is not used up by the reaction.
- Cracking involves **thermal decomposition** – using heat to break up the molecules.

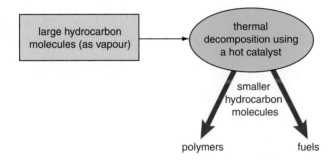

■ Cracking breaks large hydrocarbon molecules into smaller ones

Cracking produces:

- **alkanes**, which are useful as **fuels**;
- **alkenes**, which are also used as fuels and as raw materials for plastics and chemicals.

● Alkenes

- **Alkenes** have the general formula C_nH_{2n}, so ethene is C_2H_4, propene is C_3H_6 and butene is C_4H_8.
- Alkenes are **unsaturated hydrocarbons**. Unsaturated hydrocarbons contain **carbon–carbon double bonds**, shown in diagrams by C=C.
- Unsaturated hydrocarbon molecules can be drawn in the following ways:

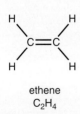
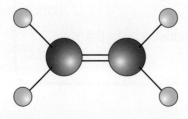

ethene
C_2H_4

FOUNDATION

It is the C=C double bond that makes alkenes reactive and useful.

The C=C double bond also makes alkenes easy to detect.

The test for alkenes

Alkenes can be detected using **bromine** or **iodine water**.

You may have seen this with ethene gas (when you cracked an **alkane**), or with a vegetable oil and even margarine.

To test for an alkene, you shake some bromine water with the suspect material in a test tube. If it does contain an alkene, then the **bromine water** is **decolorised** (the same goes for the **iodine solution**).

> **exam tip**
>
> ★ When writing about a colour change, make sure you say what the colour was to start with **and** what it was at the end. Bromine water starts brown/orange and becomes colourless. Iodine solution is either brown or purple and becomes colourless.

check your understanding

❶ Which **one** of the following sentences says the right thing about using a catalyst when cracking crude oil? *(1 mark)*

   A A catalyst is one of the reactants in the reaction.
   B A catalyst is one of the products in the reaction.
   C A catalyst speeds up a reaction but is not used up by the reaction.
   D A catalyst speeds up a reaction and is used up by the reaction.

❷ Cracking hydrocarbons involves which type of reaction? *(1 mark)*

   A oxidation
   B reduction
   C neutralisation
   D thermal decomposition

❸ Margarine will react with iodine solution, making it become colourless. This shows that margarine contains: *(1 mark)*

   A alkenes.
   B alkanes.
   C hydrocarbons.
   D a mild bleach.

C1b Oils, Earth and atmosphere

Making ethanol

Burning fuels can be 'dirty' – they often produce pollutants such as sulfur dioxide. **Ethanol** is a much 'cleaner' fuel – it burns to produce **water** and **carbon dioxide** only. There are no particulates or sulfur dioxide.

Making ethanol from ethene

- **Ethene** is an alkene produced by cracking the heavier fractions in crude oil.
- Ethene can be reacted with **steam** at high temperature and pressure with a **catalyst** to make **ethanol**.

$$C_2H_4 \ (\text{ethene}) + H_2O \ (\text{steam}) \xrightarrow[\text{high pressure}]{\text{catalyst}} C_2H_5OH \ (\text{ethanol})$$

Only 5% of the ethene is converted to ethanol, but by putting it through the reactor again and again, eventually 95% of the ethene is converted to ethanol.

The ethanol has to be separated from the water (purified) by **fractional distillation**.

Making ethanol by fermentation

- The raw material is a **starchy** plant. In hot countries such as Brazil, sugar cane is grown specially to make ethanol.
- The starch is taken out of the plants using **hot water**.
- It is then broken down into sugars by **fermentation**. This takes a few days.
- The ethanol has to be concentrated and separated from the mixture (15% ethanol) by **fractional distillation**.

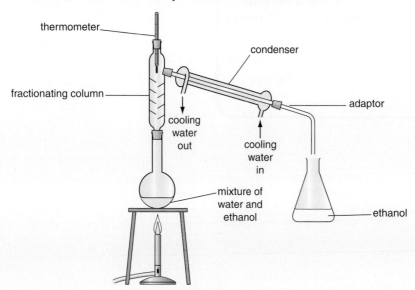

■ During fractional distillation, the vapours condense and evaporate several times in the column

Which process is best?

Is it better to use ethene or fermentation to make ethanol for use as a fuel?

Here is a table of **some** of the advantages and disadvantages of each method.

	Advantages	**Disadvantages**
Ethanol from ethene	95% conversion Simple process	Ethene comes from crude oil – a non-renewable resource High energy cost of temperature and pressure
Ethanol from fermentation	Cheap raw material Renewable plant material – doesn't use fossil fuels	15% mixture must be concentrated and purified by distillation – extra cost

If you are given extra information on making ethanol and asked to compare the methods, look at the **advantages** and **disadvantages** of each method:

- cost – which is cheaper?
- speed of production – which can be done quickly, and which do you have to wait for?
- environment – which method uses a renewable source, or uses less energy?

exam tip

★ Make two lists, headed 'Advantages' and 'Disadvantages', then make two or three relevant points under each one.

check your understanding

4 Explain why ethanol is seen as a 'clean' fuel. *(3 marks)*

5 Ethanol is purified by: *(1 mark)*

  A distillation.
  B filtration.
  C crystallisation.
  D fermentation.

6 Ethanol is produced from crude oil by: *(1 mark)*

  A fermentation.
  B catalysts.
  C cracking and distillation.
  D fermentation and distillation.

C1b Oils, Earth and atmosphere

Making polymers

- **Polymers** are really, really long molecules or chain molecules.
- **Plastics** such as **poly(ethene)** and **poly(propene)** are polymers.

Polymerisation

- Polymers can be made from **alkenes**.
- Many **small** molecules (**monomers**) join together to form very **large** molecules (**polymers**).
- For example, **ethene** molecules join up to make **poly(ethene)**, and **propene** molecules join up to make **poly(propene)**.

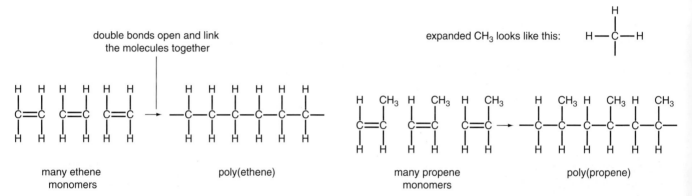

Poly(ethene) is a long-chain polymer molecule

Poly(propene) is a long-chain polymer molecule

Properties of polymers

Poly(ethene) is an example of a **thermoplastic**. Thermoplastics can easily be softened by heat and moulded into shape.

Thermosetting plastics are harder and less flexible than thermoplastics. They do not soften when heated, but set hard permanently – they can't be remoulded.

Designing different polymers

The **properties** of polymers depend on what they are made from and how they are made (conditions such as temperature, pressure, catalyst).

- For example, **slime** with different **viscosities** (runniness) can be made by adding different amounts of borax. The more borax added, the more like a solid it becomes.
- **Poly(ethene)** comes in two forms – high-density and low-density. They are made at different pressures with different catalysts.
- **Poly(chloroethene)** – that's PVC – is rigid, but can be modified to make softer, more flexible forms by adding small 'plasticiser' molecules to the polymer.

FOUNDATION

Here are some examples of new polymers made with special properties:

- **smart polymers** respond to changes in their surroundings – smart packaging materials could change colour when the temperature rises.
- **hydrogels** are polymer gels that grow or shrink when water is added – they are used in wound dressings and nappies.
- **shape-memory polymers** can stretch, but return to their original shape when heated.
- waterproof materials such as Gore-Tex® have tiny holes that let water vapour (sweat) through, but not liquid water.

■ The low-density poly(ethene) bottle will float, even when full of water, but the high-density lid sinks – try it at home

Crude oil – too valuable to burn?

Some people think the limited supply of crude oil should be kept for making **useful products** instead of burning it to give us **energy**.

Can you evaluate (work out) the social and economic advantages and disadvantages of using products from **crude oil** as **fuels** or as **raw materials** for plastic? There's more about this in the next section.

check your understanding

7 Name the polymer that can be made from propene. *(1 mark)*

8 Insert these words in the numbered gaps in the following: *(4 marks)*

    large   monomers   ethene   alkenes

Polymers can be made from ____1____. In these reactions, many small molecules (____2____) join together to form very ____3____ molecules (polymers).

For example, ____4____ molecules join up to make poly(ethene).

9 The properties of polymers depend on (choose one): *(1 mark)*

  A their colour and shape.
  B when they were invented.
  C how they were made.
  D what they are used for.

exam tip

★ Don't forget to cross out the wrong answers as you find them.

C1b Oils, Earth and atmosphere

Waste-disposal problems

Most plastics are not **biodegradable**, so they are not rotted (**decomposed**) by microorganisms. This can lead to too much waste.

One solution would be to make **biodegradable** carrier bags and food wrapping. But the new polymer has to last as long as it is needed, before it rots away.

● What happens to non-biodegradable polymer waste?

Polymer waste can be disposed of in three ways:

- landfill;
- recycling;
- incinerator.

● Does it matter anyway?

Let's try to **evaluate** (work out) the effects of:

– using and disposing of polymers
– recycling polymers.

Here's some information for you to consider:

- about 500 billion to 1 trillion plastic bags are used and discarded annually worldwide – more than a million per minute
- in New York City, one less bag per person per year would save US$250 000 in disposal costs
- when 1 ton of plastic bags is re-used or recycled, the energy equivalent of 11 barrels of oil is saved.

Tables are useful when it comes to looking at a lot of information together. The one on page 65 has just a few ideas for you. You may have some of your own that are just as good.

It is useful to learn the **evaluation** technique:

- what are the pros and cons?
- use the information provided, but check for people who might be biased.
- use the evidence of data in graphs or tables to support your argument.
- come to your conclusion.

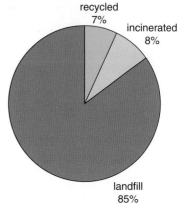

■ Most plastic waste ends up in the ground

exam tip

★ If you're asked to **evaluate** something, you must remember to give a **balanced** argument and deal with **the evidence for and against** – forget one and you'll lose marks.

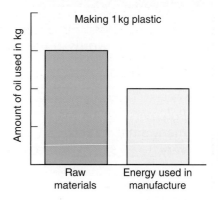

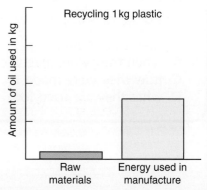

■ Recycling plastic saves oil resources, both as raw materials and as energy used in manufacturing

FOUNDATION

	Effects of using, disposal of or recycling polymers		
	Social (people)	Economic (business)	Environmental (nature)
Uses of polymers	Many things that people use are made of polymers, such as sports equipment, clothing, food packaging, ICT and high-tech items	Creates employment in manufacturing and designing products	Uses fossil fuel (oil) as raw material Making the polymer may produce toxic waste Creates non-biodegradable waste
Disposal of polymers	It's smelly and looks horrible if you live close to a landfill site	Creates employment: – refuse collectors – landfill/incinerator workers	Uses up land The landfill is unsightly Incineration produces a lot of ash, greenhouse gases and toxic fumes, but it can also be used to generate energy
Recycling of polymers	Happy communities because their environment is cleaner But there is effort involved in individuals having to sort their rubbish	Creates employment Is recycled plastic more expensive than new plastic?	Saves crude oil Produces much less pollution Reduces use of landfill sites Uses energy, possibly more than the energy needed to make new plastic

check your understanding

10 Recycling supermarket carrier bags is good for the environment because: *(1 mark)*

  A it saves money.   B it saves crude oil.
  C it saves arguing about it.   D it saves lives.

11 Here is a table showing how you could start to work out some of the pros and cons of using products from crude oil, either as fuels or as raw materials for plastic. Put one valid comment in each empty box. *(4 marks)*

	Advantages	Disadvantages
Crude oil products as fuels		
Crude oil products as raw materials		

FOUNDATION

C1b Oils, Earth and atmosphere

Vegetable oils and fuels

Vegetable oils are important **foods** and **fuels**; they provide **energy**.

● Vegetable oils in food

- Vegetable oils are important foods. As well as providing **energy**, they also give us **nutrients**, especially vitamin E.
- Frying food in oil adds flavour and enables a high temperature to be reached – food cooks more quickly.
- Vegetable oils are found in some fruits, seeds and nuts.
- They are **extracted** by crushing the plant material and removing the oil by **pressing** or **distillation**.
- Vegetable oils that are **unsaturated** contain carbon–carbon double bonds. These can be found by reacting with **bromine** or **iodine water**.

Some common plant oils and the sources of the oil	
Plant oil	Where oil is stored in the plant
Olive oil	Fruit
Rape oil	Seeds
Peanut oil	Nut (food store for seeds)
Avocado oil	Fruit
Sunflower oil	Seed
Palm oil	Fruit

● Turning plant oils into margarine

- Vegetable oils are usually runny, and can't be used to make cakes or pastry.
- They can be hardened by reacting them with hydrogen and a nickel catalyst at about 60 °C. This process is called **hydrogenation**.
- Making margarine by hydrogenation turns **unsaturated** fats into **saturated** fats.
- The hydrogenated oils (margarine) have higher melting points, so they are solids at room temperature.

Some people say that margarine is better for you than butter.

Can you evaluate the effects of using vegetable oils in foods, and the impacts on diet and health? Compare the **total** amount of saturated fat in 100 g margarine and butter.

Remember that:

- too much dietary fat and not enough exercise can lead to obesity;
- saturated fats can lead to heart disease and stroke.

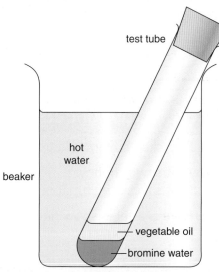

■ The bromine water test for unsaturation

■ Hydrogenation

Comparing butter and different margarines			
	Butter	Hard margarine	Soft margarine made with olive oil
Fat per 100 g in g	81	81	59
Of which saturates in %	64	20	24
Cholesterol per serving in mg	30	0	0

FOUNDATION

Vegetable oils for fuels – biodiesel

Some **biodiesel** is made from used vegetable oil. Most biodiesel is made from crops. Biodiesel is renewable. We don't use up crude oil.

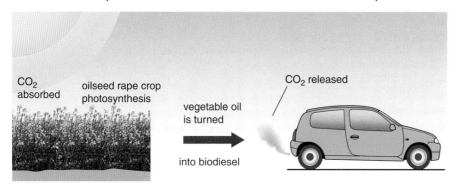

■ Burning biodiesel produces no CO_2 overall; CO_2 is absorbed by the plants from which the fuel is made

As well as benefits, there are drawbacks of using vegetable oils to produce fuels, as shown in the table.

Benefits and drawbacks of using growing crops to produce biodiesel	
Benefits	Drawbacks
No modification of diesel engine needed	A lot of land used to grow plants for fuel instead of food
Can be mixed with ordinary diesel	
Much **less polluting** than ordinary diesel	
It is **carbon-neutral**	
Biodiesel can be produced locally – local economic benefit	

exam tip

★ You don't need to remember the summary table – but you may be asked to analyse similar information in the exam and sort it out into lists like these.

check your understanding

12 a) What contains most unsaturated oil – butter or olive oil? *(1 mark)*
  b) Explain why unsaturated fats are thought to be healthier than saturated fats. *(2 marks)*
  c) Give the steps needed to turn an unsaturated fat into a form useful for baking. *(3 marks)*

13 What is most biodiesel made from? *(1 mark)*

14 The chemical reaction to form solid fats from liquid oils is called: *(1 mark)*

  A neutralisation.   B freezing.   C cracking.   D hydrogenation.

15 Biodiesel is carbon-neutral – that means: *(1 mark)*

  A it does not produce soot when it burns.
  B the exhaust gases contain no carbon.
  C the exhaust gases are neutral.
  D it produces the same amount of CO_2 as was absorbed by the plants from which it was made.

FOUNDATION

C1b Oils, Earth and atmosphere

Food additives and emulsifiers

● **Do you know what you're eating?**

Colours, flavours and stability are often changed by processing food.

Processed foods may contain **additives** to improve:

- appearance – imagine colourless strawberry ice cream;
- taste – we all add a little salt to our chips;
- shelf-life – to make it last longer.

The European Union says that additives must be listed in the ingredients, so we know what we are eating.

Some additives have been given **E-numbers** to show they have passed safety tests.

Even natural substances used as additives have E-numbers: vitamin C is E300.

● **Emulsifiers in food**

- **Oils** do not **dissolve** in water, but they can be used to produce **emulsions**.
- **Emulsions** are special **mixtures** of oil and water. They are thicker than oil or water, and have many uses that depend on their special properties.
- Emulsions taste better in the mouth, stick to the food and look more appetising than oil or water separately. Mayonnaise, salad cream and ice creams are emulsions.
- As oil and water do not mix, emulsions have to be 'persuaded' to stay together. **Emulsifiers** do this. French salad dressing (olive oil and vinegar) has no emulsifier and you can watch the two separate just after you've shaken them up.

● **Are food additives safe to eat?**

E-numbers are supposed to be safe in food. But many people (maybe a friend, or yourself) are allergic to a **colouring**, **flavouring** or **preservative**.

Remember: there are **benefits** in using food additives, but there are also **drawbacks** and **risks** involved.

■ Emulsifiers keep emulsions emulsified

exam tip

★ Another chance to **evaluate** – remember that both sides must be mentioned, e.g. additives prevent food spoiling but may cause poor health in some people.

Food analysis

- Clever chemists have many ways of finding out exactly what is in food, and how much of it there is.
- One way of detecting **artificial colours** is **chromatography**. You may have done this with the tips of felt pens, food colourings or even the colours from Smarties™.

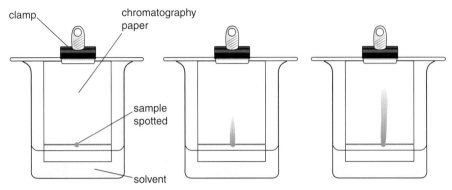

■ Simple chromatography

In laboratories, a more **sensitive**, **reliable** and **accurate** type of **chromatography** is used.

check your understanding

16 If a food additive has an E-number, that means it is: *(1 mark)*

   A edible.
   B allowed to be used.
   C English.
   D an even number.

17 Chromatography can be used to detect: *(1 mark)*

   A colourings.
   B flavourings.
   C preservatives.
   D emulsifiers.

18 Emulsions without emulsifiers: *(1 mark)*

   A are not real emulsions.
   B separate quickly.
   C go bad too soon.
   D have no oil in them.

FOUNDATION

C1b Oils, Earth and atmosphere

The Earth and continental drift

● The structure of the Earth

- The Earth is made up of a **core**, **mantle** and **crust**.
- The **core** has two parts. The inner core is solid, the outer core is liquid.
- The **mantle** is a thick band of mostly solid rock, but it can flow very slowly.
- The **crust** is really quite thin.

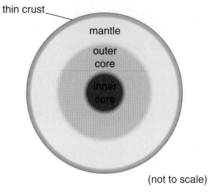

■ A section through the Earth

A long time ago, scientists thought that:

- as the Earth had cooled down, it got smaller;
- this caused wrinkling on the surface;
- this wrinkling is seen as mountains, valleys, volcanoes, etc.

> **exam tip**
> ★ Learn these three points to explain the old theory.

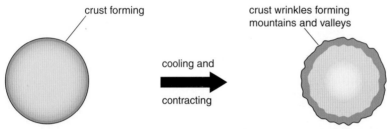

The Earth 4000 million years ago.

The Earth after cooling for millions of years.

■ An early idea about how mountains and valleys formed on the Earth's surface

Then a weatherman named Wegener came up with a totally different idea.

● Continental drift

Wegener saw (observed) that on opposite sides of the Atlantic Ocean, there were:

- rocks of the same age;
- the same fossils;
- coastlines that seemed to fit together.

He said this could be explained if:

- all the continents were once joined together as one supercontinent;
- the supercontinent broke up and the continents had slowly moved apart.

Wegener used the 'jigsaw fit' and similar rock types and fossils of South America and Africa as evidence that the continents had moved.

But the theory of **continental drift** was not accepted by scientists for many years.

FOUNDATION

- Some were hostile – after all, the theory came from a weatherman, not a geologist.
- Some scientists said a bridge of land between the continents could have allowed animals to walk across.
- Nobody knew what was at the bottom of the oceans between the continents.

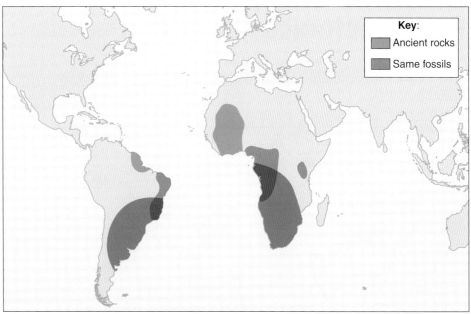

■ Wegener's evidence for the movement of South America and Africa

Today, we accept Wegener's theory. This is an example of how scientific ideas can change with time as scientists check each other's ideas and evidence. There is now new evidence about the ocean floor.

check your understanding

19 Draw a labelled diagram clearly showing the structure of the Earth. Use the labels:

        mantle   inner core   outer core   crust   (4 marks)

20 The mountains and valleys on the surface of the Earth were once thought to be the result of:

                                                                                                (1 mark)

  A the Earth becoming wrinkly as it dried out.
  B the Earth cracking with too much heat.
  C the Earth cooling and shrinking.
  D giant creatures in the core pushing up the rocks.

21 Look carefully at these two statements: do they support or contradict Wegener's theory?

                                                                                                (1 mark)

  A Different continents have the same fossils.
  B Similar ancient rocks and mountain chains appear on different continents.

22 A new scientific theory becomes accepted by other scientists: (1 mark)

  A when they read it.
  B as they check the evidence for themselves.
  C when they see it on the TV news.
  D when they hear it on the radio.

FOUNDATION

C1b Oils, Earth and atmosphere

Plate tectonics

The theory that replaced continental drift is called the **theory of plate tectonics**.

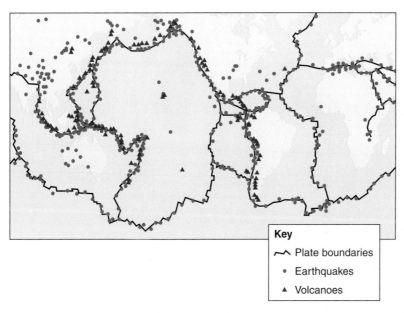

■ The tectonic plates

This theory says:

- The Earth's **crust** and the top of the **mantle** are cracked into large pieces called **tectonic plates**.
- **Radioactivity** deep inside the Earth produces heat – just like in nuclear reactors.
- This heat causes **convection currents** in the mantle, which make the **tectonic plates** move.

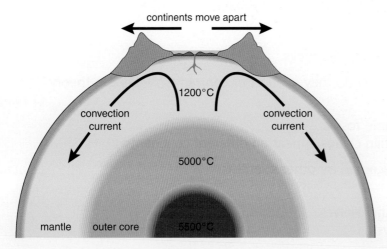

■ Convection currents make tectonic plates move

72 FOUNDATION

Earthquakes and volcanoes

- Usually the **tectonic plates** move extremely slowly – only a few centimetres a year. The movement causes stresses and strains to build up between the plates.
- These stresses and strains can be released when the plates suddenly move.
- It's at the boundaries between **tectonic plates** that earthquakes and volcanic eruptions occur.

Where, but not when

- Scientists can predict **where** earthquakes and volcanic eruptions will occur.
- But they cannot predict **when** they will happen.

They can measure:

- the forces in rocks;
- tiny movements and bulges in the Earth's crust;
- shock waves from small earthquakes that happen before the main one;
- movement of the tectonic plates using global positioning systems (GPS).

These may give a **warning sign** that an earthquake or eruption is **likely**, but not a definite prediction of when.

check your understanding

㉓ Tectonic plates are in which part of the Earth? *(1 mark)*

  A The crust and mantle.
  B The mantle and core.
  C The outer core.
  D The inner core.

㉔ Tectonic plates move because of: *(1 mark)*

  A conduction.
  B convection.
  C radiation.
  D pressure.

㉕ The heat causing plate movement is generated by: *(1 mark)*

  A burning fossils.
  B burning fossil fuels in the Earth's core.
  C radioactive processes in the Earth's core.
  D solar power.

㉖ Where and why do earthquakes occur? *(2 marks)*

C1b Oils, Earth and atmosphere

Gases in the atmosphere

● Nitrogen and oxygen

For 200 million years, the gases in the **atmosphere** have been more-or-less like they are now:

- about four-fifths (80%) **nitrogen**;
- about one-fifth (20%) **oxygen**;
- small amounts of other gases, including **carbon dioxide**, **water vapour** and **noble gases**.

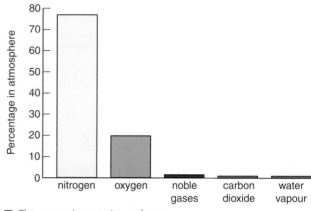

■ The gases in our atmosphere

● The noble gases

(Not posh – just unreactive!)

The **noble gases** are in Group 0 of the periodic table. That's the column on the right with helium at the top.

Noble gases are really useful in:

- **filament lamps** ('normal' light bulbs);
- **electric discharge tubes** (neon lights).

That's because they are all unreactive. For example, the hot filament of a light bulb burns up in less than one second in air, but stays bright for ages in argon – it does not react with it.

Helium is much less dense than air. Therefore it is used in balloons – for parties, carnivals, and even scientific experiments high in the atmosphere.

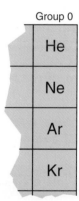

■ The position of the noble gases in the Periodic Table

■ Helium balloons rise because they are less dense than air

exam tip

★ Remember to say what you mean: a lot of people say that helium is 'lighter than air', but a tonne of helium weighs the same as a tonne of air. You must say it is 'less dense than air'.

FOUNDATION

Carbon dioxide in the atmosphere today

- There are processes that add **carbon dioxide** (CO_2) to the atmosphere, and processes that take it away.
- However, there is an increase in the level of CO_2 in the atmosphere.
- Burning **fossil fuels** (to produce energy) adds CO_2 to the atmosphere.
- There is a **correlation** between the sudden increase in CO_2 levels and the beginning of the Industrial Revolution.
- This increase in CO_2 is thought to be causing global warming.

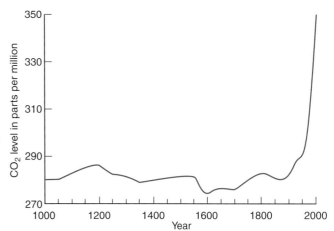

■ The amount of carbon dioxide in the atmosphere has increased rapidly since about 1860

check your understanding

27 The proportions of nitrogen and oxygen in the atmosphere are: *(1 mark)*

  A 20% and 80%.
  B 30% and 70%.
  C 70% and 30%.
  D 80% and 20%.

28 Argon is used in filament lamps because: *(1 mark)*

  A it is cheap.
  B it is colourless.
  C it does not react with the filament.
  D it does not matter what is used.

29 Helium is used in balloons because: *(1 mark)*

  A it is less dense than air.
  B it is heavier than air.
  C it is cheaper than hydrogen.
  D there is not enough nitrogen.

FOUNDATION

C1b Oils, Earth and atmosphere

Theories about the atmosphere

● Where did it come from?

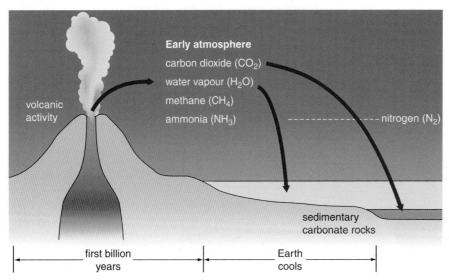

■ The Earth's early atmosphere came from volcanoes

During the first billion years of Earth, there was an enormous amount of volcanic activity.

Volcanoes gave out the gases that made up the early atmosphere:

- carbon dioxide (CO_2) – main part of the early atmosphere;
- methane (CH_4) – small part;
- ammonia (NH_3) – small part;
- water vapour (H_2O).

This water vapour condensed to form the oceans.

● Alternative theories about the atmosphere

There are different theories about the proportions of these gases in the early atmosphere. These theories are hard to test, as there is no direct evidence.

Some theories suggest that the source of water was **comets**, not volcanoes.

Explaining the change in carbon dioxide and oxygen levels

- Notice that the atmosphere started as more-or-less only **carbon dioxide** with almost no **oxygen**.
- This would have been like the atmospheres of **Mars** and **Venus** today.
- As plants evolved, they replaced **carbon dioxide** with **oxygen** by **photosynthesis**.
- Carbon from **carbon dioxide** in the air became locked up in rocks and fossil fuels.
- This gave us our atmosphere of today.

What's happening in the twenty-first century?

- Sea water removes CO_2 from the atmosphere (by absorbing it).
- This counteracts the effect of burning fossil fuels – but only a little.
- Plants (via photosynthesis) also remove CO_2 from the atmosphere.

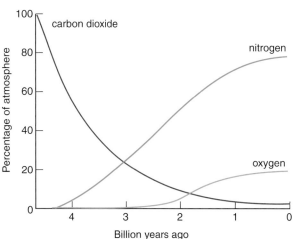

■ How the Earth's atmosphere has changed

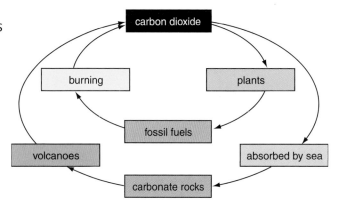

■ Carbon dioxide moves into and out of the atmosphere

check your understanding

30 The early atmosphere was like the current one of which two planets? *(1 mark)*

   A Mercury and Mars   B Mercury and Venus
   C Mars and Venus   D Venus and Saturn

31 Burning fossil fuels is a major human activity. We burn fossil fuel because: *(1 mark)*

   A it makes a nice fire to sit round.   B it is old and needs using up.
   C the fossils are not needed any more.   D we get energy from them.

32 Carbon dioxide leaves the atmosphere when (choose **two**): *(2 marks)*

   A it snows.   B plants use it for photosynthesis.
   C volcanoes erupt.   D the sea absorbs it.

FOUNDATION

P1a Energy and electricity

Thermal radiation

Infra-red and thermal radiation

- An electric fire **emits** (gives out) heat. The heat is transferred from the fire as **infra-red** waves. When you sit in front of the fire, you feel warm because you **absorb** (take in) some of the heat.
- Objects that absorb (take in) heat faster than they emit it (give it out) will warm up (**increase in temperature**).
- All objects emit **infra-red radiation**. Infra-red radiation is also called **thermal radiation**.

Everything absorbs and emits thermal radiation

But some objects are better absorbers and emitters than others. The chart gives the difference between surfaces.

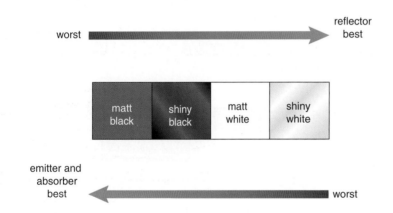

■ Shiny aluminium foil helps to keep food hot for longer by reflecting thermal radiation back into the food

- Dark surfaces are better absorbers and emitters of thermal radiation than light surfaces.
- Shiny surfaces reflect thermal radiation better than matt (dull) surfaces.
- **Shiny surfaces reduce heat transfer by reflecting thermal radiation.**

Finding out how the surface affects absorption of thermal radiation

The picture shows a simple way of comparing different surfaces.

- If many people do the same experiment, the results would be the same; so the results are **reliable**.
- Not reading the thermometer correctly could give an **anomalous** result. The anomalous result will not fit the pattern.

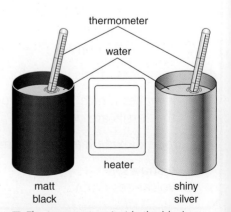

■ The temperature inside the black container goes up faster than the temperature inside the shiny silver container

- Using a temperature sensor and data logger instead of a thermometer would make the temperature measurements more **accurate**. This is because you are less likely to read the wrong temperature.
- Increasing the accuracy means that the temperatures written down are closer to the real temperatures.
- A temperature sensor is probably also more **precise** – it can measure smaller changes.

exam tip

★ These words are used a lot in investigations. Make sure you remember what they mean.

check your understanding

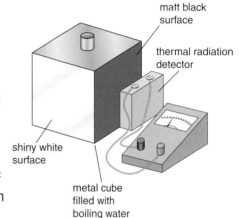

1 Match each word (A–D) to one of the numbers (1–4) in the text: *(4 marks)*

A emit   B white   C cooler   D absorb

In hot countries, houses are often painted ____1____. During the daytime, the walls slowly ____2____ thermal radiation. At night, the walls slowly ____3____ thermal radiation. This keeps the house ____4____ in the daytime and warmer at night.

2 A hollow metal cube is filled with boiling water. Each side of the cube is different. The values in the table were taken with a thermal radiation detector facing each side of the cube. The higher the value, the greater is the amount of thermal radiation emitted by the surface.

Match each side (A–D) to one of the surfaces listed below: *(4 marks)*

matt black   matt white   shiny black   shiny silver

Side	Detector value
A	0.37
B	0.72
C	0.76
D	0.26

3 The graph shows the temperature inside a room.
 a) At which point (X, Y or Z) is the room temperature constant? *(1 mark)*
 b) At which point (X, Y or Z) is the room absorbing and emitting the same amount of thermal radiation each second? *(1 mark)*
 c) The temperature of the room goes up at the same steady rate for another hour. What is the room temperature at 13.00 h? *(1 mark)*

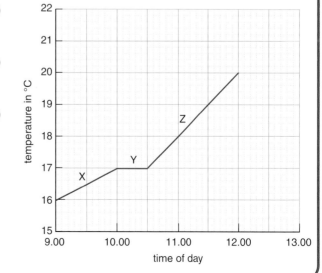

exam tip

★ You need to **extrapolate** the graph. This means make the line longer, but follow the same pattern.

FOUNDATION

P1a Energy and electricity

Conduction and convection

Heat transfer by **conduction** and **convection** involves the movement of particles.

> **exam tip**
> ★ Conduction and convection are almost the same words, but mean very different things. Make sure you know which one an exam question is asking about.

● Heat travels through solids by conduction

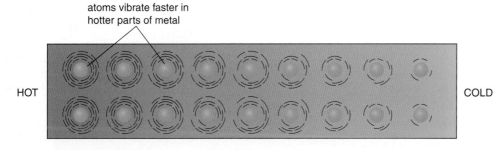

- The particles at the hot end gain energy.
- The particles **vibrate** faster and take up a bigger space.
- The particles bump into other particles nearby.
- The energy is passed from one particle to another.
- So the energy has been conducted through the metal.

Metals are better **conductors** than non-metals. A poor conductor is a good **insulator**.

● Air is a good insulator

A polar bear has a thick fur coat. Air is trapped in the fur. The air reduces the energy lost by conduction. This helps to keep the polar bear warm.

● Heat travels through liquids and gases by convection

To transfer heat by convection, a liquid or a gas must move. This movement is called a **convection current**.

- Air particles near the heater gain energy and move around more quickly.
- The particles move apart and take up more space.
- The warm air **expands** and rises.
- The warm air rising pushes heavier, colder air downwards.

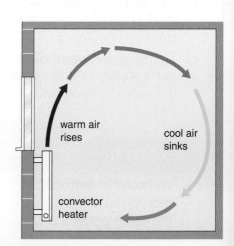

FOUNDATION

> **exam tip**
> ★ To explain how convection happens in water, just change the word 'air' to 'water'. Don't learn a whole new explanation.

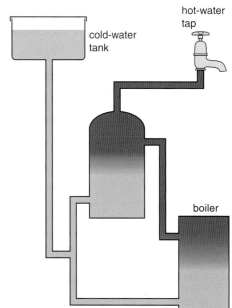

Hot-water systems work by convection

- A boiler heats the water.
- The hot water rises to the hot-water tank.
- Colder, heavier water falls from the cold-water tank.
- A convection current keeps the water moving.

Question: 'When is a radiator not a radiator?'

Answer: 'When it's a radiator in a central heating system.'

The radiators in a central heating system are really convector heaters. They heat up the air, creating a convection current that moves the heat around the room.

check your understanding

4 Which **two** statements describe the transfer of heat through water? *(2 marks)*

A Heat is transferred by electromagnetic waves.
B Heat transfer involves particles.
C Heat is transferred because the water expands.
D Most heat is transferred by conduction.

> **exam tip**
> ★ Write out the two correct statements. It will help you remember them.

5 Which **one** of the following statements is true? *(1 mark)*

A Heat always moves from a cold object to a hot object.
B Heat is transferred through a gas by conduction.
C Heat can be transferred through a solid by convection.
D Expanded foam is a good insulator because it traps air.

6 Explain why the ice box is always at the top of a fridge. *(2 marks)*

7 Complete the following sentences, using the words below: *(4 marks)*

conductors   insulators   liquid   particles

Metals are good ____1____. Plastic and glass are poor conductors so they are good ____2____. Without ____3____, heat could not travel by convection through a ____4____.

FOUNDATION 81

P1a Energy and electricity

Reducing rates of heat transfer

● Some things transfer heat more quickly than others

How fast (the **rate**) something transfers heat depends on:

- what it is made from;
- its shape;
- its size;
- the difference between its temperature and the air around it.

Comparing results	What we have found out
A and C or A and B	The bigger the temperature difference between the water and the air, the faster the water loses heat
A and C	A metal mug conducts heat away from the water more quickly than a ceramic mug
A and B	The bigger the surface area, the more quickly the water loses heat

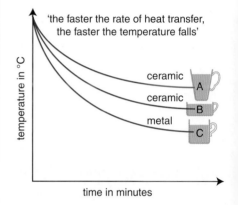

The comparison of these graphs is a **fair test** because:

- only one variable was changed – the shape of the mug, or what the mug was made from (the **independent variable**);
- the volume of the water was always the same and the water always started at the same temperature (**control variables**).

● A vacuum flask keeps hot drinks hot and cold drinks cold

Heat is transferred by conduction, convection and radiation.

The heat transfer from a hot drink inside a vacuum flask to the colder air outside is **reduced** by:

- the **vacuum**, which stops **conduction** and **convection** out of the sides;
- the **silver coating**, which reflects heat, reducing heat transfer by **radiation**;
- the plastic stopper, which is a good insulator.

The flask keeps an iced drink cool because the same features reduce the heat entering the flask.

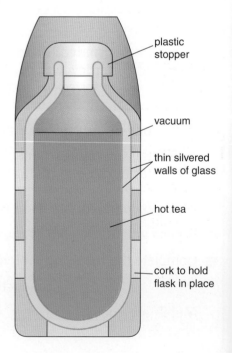

82 FOUNDATION

Keeping your home warm is about reducing heat loss

Most types of **insulation** involve **air**.

- Air trapped in small air bubbles cannot move far. If the air can't move far, then the heat lost by convection is reduced.
- Air is a good insulator, so the heat lost by conduction is reduced.

Shiny materials are often used in the loft or behind radiators to reflect heat. This reduces the heat lost by radiation.

Draught excluders help keep a house warm by stopping cold air entering the house.

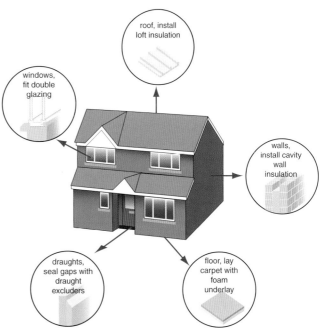

■ The heat lost through the roof, walls, windows, floor and doors of a house can be reduced by using insulation

exam tip

★ An exam question may not be about house insulation, but if it's about keeping warm, the answer is probably about trapped air reducing heat loss by convection and conduction.

check your understanding

8 The diagram shows a model solar water heater made by a student.

Complete the following sentences by drawing a ring round the correct word in each case.
   a) Copper is a good **insulator** / **conductor** / **radiator** of heat. (1 mark)
   b) Painting the pipe black **decreases** / **increases** the rate at which the pipe absorbs heat from the Sun. (1 mark)
   c) Heat lost by **conduction** / **convection** / **radiation** is reduced by the aluminium foil. (1 mark)
   d) The foam is a good insulator because it traps **air** / **water** / **light**. (1 mark)

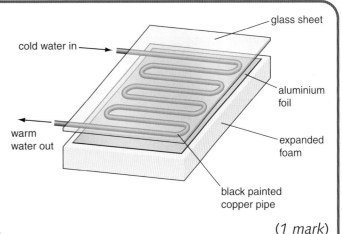

9 Which statement gives a reason why woollen gloves keep your hands warm? (1 mark)

   A They are tight fitting.
   B Air is trapped in the woollen fibres.
   C They can be bought in different colours.
   D Wool is a natural material.

FOUNDATION 83

P1a   Energy and electricity

Energy efficiency

● **You can't get energy from nothing – it's just impossible!**

Energy can be:

- moved from one place to another – this is **transferring** energy;
- changed from one form into another – this is **transforming** energy.

Remember that you cannot make energy and you cannot destroy energy.

● **Energy makes things work, but only when it's transformed**

A car engine transforms the chemical energy stored in the fuel to kinetic energy, heat and sound.

The energy transformation can be shown in a **Sankey diagram**. The wider the arrow, the bigger is the fraction of energy transformed to that form.

The Sankey diagram shows that the **energy input** from the fuel and the total **energy output** from the engine are the same. But the heat and light are not really wanted – this is **wasted** energy. The more energy the engine wastes, the less **efficient** it is.

An **efficient** device is good at transferring input energy into useful output energy.

$$\text{efficiency} = \frac{\text{useful energy transferred by the device}}{\text{total energy supplied to the device}}$$

- Energy is measured in **joules (J)**.
- Efficiency doesn't have a unit, you just write it as a decimal number.

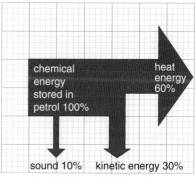

■ Sankey diagram for a petrol engine

> **exam tip**
>
> ★ You do not need to remember this equation, but you do need to be able to use it. The number on the top of the equation is always smaller than the number on the bottom. This means the answer is always less than 1.

● **Energy can never vanish**

- All energy, useful and wasted, goes into the surroundings, which become warmer.
- The energy becomes spread out and difficult to use for any more energy transformations.
- The energy has not gone – it's just not useful any more.

Sometimes the energy that is normally wasted can be usefully used. A car heater warms the inside of a car with waste heat from the engine.

Payback time and cost effectiveness are not the same

Payback time is how long it takes to get back the money spent on reducing the amount of energy used.

Example: spending £240 on loft insulation saves £80 each year on energy bills.

Payback time = 240/80 = 3 years.

How **cost effective** it is to buy something that reduces the amount of energy used depends on:

- how much you spend on buying and putting it in (initial cost);
- how often it has to be replaced;
- how much money is saved each year on energy bills.

If the loft insulation lasts 25 years before it needs replacing, and energy costs stay the same, the total amount of money saved on the energy bills will be:

25 × £80 = £2000

The insulation cost only £240, so over 25 years the total saving will be:

£2000 − £240 = £1760

That's a lot of money to save. The insulation is very cost effective.

check your understanding

10 The table compares different ways of reducing the heat loss from a house.

Calculate the payback time for each method of reducing heat loss.
(3 marks)

Method of reducing heat loss	Initial cost in £	Money saved each year on energy bills in £
Carpets	1500	75
Double glazing	5600	140
New hot water tank	150	30

11 The energy input to a solar cell is 2500 J every second. The useful energy output is 200 J per second.
a) How much energy does the solar cell waste every second? *(1 mark)*
b) Which of answers A–D gives the efficiency of the solar cell? *(1 mark)*

A 12.5   B 11.5   C 0.8   D 0.08

12 A new freezer uses less energy than an old freezer to keep the same amount of food frozen. This means that, compared with the new freezer, the old freezer: *(1 mark)*

A costs less per day to run.   B is more efficient.
C is less efficient.   D wastes less energy.

FOUNDATION

P1a Energy and electricity

Electrical power and energy costs

Electrical energy is easily transformed into other forms of energy. This is what makes it so useful.

Power is how fast energy is transformed.

It is measured in **watts** (W).

- 1 kilowatt (kW) = 1000 watts (W).
- A device that transforms 1 joule of energy every second has a power of 1 watt.

The total electrical energy transformed into other forms of energy by a device depends on:

- how long the device is used for;
- the power of the device.

Cost depends on the amount of energy transferred from the mains supply.

The cost of using a device: It is calculated using these **two equations**.

energy transferred = power × time
(in **kilowatt-hours**, kWh)　(in kilowatts, kW)　(in hours, h)

total cost = number of kilowatt-hours × cost per kilowatt-hour

This is the **only** time energy is calculated in kilowatt-hours.

Example: calculate the cost of using a 2 kW fire for 3 h. One kilowatt-hour of energy costs 12 p.

energy transferred = 2 × 3 = 6 kWh

cost = 6 × 12 = 72 p

Working out the electricity bill

The readings on the **electricity meter** are used to calculate the electricity bill. The readings are in **kilowatt-hours**.

First: subtract the two meter readings. 57139 − 56309 = 830

This is the number of kilowatt-hours of electrical energy used.

Second: multiply the number of kilowatt-hours by how much 1 kWh costs.

830 × 12 p = 9960 p = £99.60

This is the cost of the electrical energy supplied between 9 February and 11 May.

Devices using electricity	Useful energy output from device
Loudspeaker, bell	Sound
Grill, toaster, fire, iron, kettle	Heat
Lamp, computer screen	Light
Motor	Kinetic

exam tips

★ Think about what you use a device for, and the useful energy output should be obvious. You listen to a radio, so the useful output must be sound.

★ These equations will be given in an exam question. But to get the right answer, you must make sure power is in kilowatts and time in hours.

February　May

■ An electricity meter records the energy supplied in kilowatt-hours

■ The readings on the electricity meter are used to calculate the total energy cost over 3 months

FOUNDATION

Choosing the right device for the job

Different devices that do the same job have advantages and disadvantages.

For example, there are many different types of heater. Which would you choose?

For a quick blast of heat in a garage or kitchen, the fan heater would probably be best. To keep the chill out of a bedroom, it's probably the oil-filled radiator.

Heater type	Advantages	Disadvantages
2.5 kW fan heater	Two heat settings Quickly warms up a room Cheap to buy Small	Expensive to run Noisy Parts of a room are at different temperatures
0.4 kW oil-filled radiator	Cheap to run Makes no noise All parts of a room are at the same temperature	Only one heat setting Takes a long time to warm a room up Expensive to buy Big and bulky

exam tip

★ If you are asked to compare different devices, describe similarities and differences. Often when you compare devices that do the same job, an advantage for one device is a disadvantage for the other.

check your understanding

13 How much energy does a 60 W filament lamp switched on for 10 seconds transfer? *(1 mark)*

A 60 J   B 600 J   C 3600 J   D 36 000 J

14 An electricity meter reads 35 270 at 9.00 am. By 9.00 pm the same day, the following appliances have been used:

3 kW oven for 2 hours;
2 kW dishwasher for 1 hour;
2 kW fire for 6 hours.

What is the reading on the electricity meter at 9.00 pm? *(1 mark)*

A 35 250   B 35 260   C 35 280   D 35 290

15 The data in the table compare a filament lamp with a low-energy lamp. Both lamps give out the same amount of light.

Calculate the total saving made over 5 years by replacing one filament lamp with a low-energy lamp. *(3 marks)*

Type of bulb	Cost to buy in £	Number of bulbs used in 5 years	Cost of energy used in 5 years in £
Filament	0.70	5	13
Low-energy	2.50	1	2.60

exam tip

★ Always remember to include the cost of buying the appliance.

FOUNDATION

P1a Energy and electricity

Generating electricity and the National Grid

Generating electricity is about transforming energy

Most electricity is generated in power stations using **non-renewable fuels**.

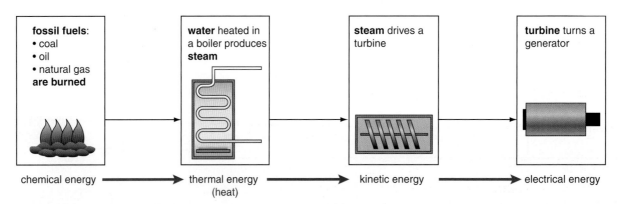

- Non-renewable fuels are coal, oil, natural gas or nuclear.
- Once non-renewable fuels are gone, they are gone forever.
- Modern gas power stations use the heat from the burning gas to heat air. The fast-moving air drives the turbine. Because they do not need to boil water, gas power stations are quick to start generating electricity.
- Nuclear power stations do not burn fuel. Heat is given out by **nuclear fission** reactions in the uranium or plutonium fuel. The heat is used to boil water and produce steam.

Different types of power station have different start-up times. **Start-up time** is how long it takes a power station to begin to generate power from the fuel. It could be a few days for a nuclear power station.

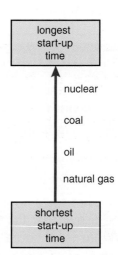

The National Grid

The **National Grid** is a network of cables and transformers for **transferring** energy. It links the power stations that generate electricity to the people (consumers) who use it.

FOUNDATION

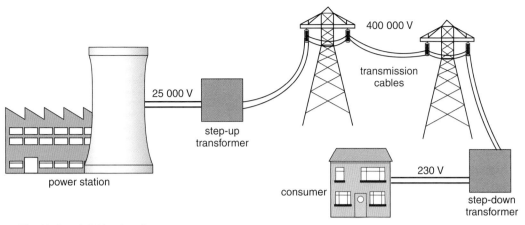

■ The National Grid network

Transferring electrical energy at a high voltage increases efficiency

- The power station generates electrical energy.
- The electrical energy is transferred to a **step-up transformer**.
- The transformer increases the **voltage (potential difference)** across the transmission cables.
- This reduces the **current** through the cables.
- The smaller the current, the less electrical energy is transformed into heat and **wasted**.
- Wasting less energy makes the transfer of electrical energy more **efficient**.

At the end of the transmission cables, the electrical energy is transferred to a series of **step-down transformers**. These transformers decrease the voltage (potential difference) to a value safe for you to use.

exam tip
★ You do not need to know how transformers work, only why they are used.

check your understanding

16 Which type of power station is the quickest to start generating electrical energy when restarted after being closed down for maintenance? *(1 mark)*

A nuclear   B natural gas   C coal   D oil

17 How is the electricity generation process in a nuclear power station different from that in a coal-burning power station? *(1 mark)*

18 It takes 5 years to build a new nuclear power station. Each year, the power used in construction is 250 MW.
   a) How much power is used to build the power station? *(1 mark)*

   Once the power station is working, it generates 1000 MW of power each year.
   b) How long will it be from the time when it starts working until the power station has generated more electrical energy than the energy used to build it? *(1 mark)*

FOUNDATION

P1a Energy and electricity

Renewable energy resources

- **Renewable energy resources** are replaced as quickly as they are used.
- Most renewable energy resources used to generate electricity do not burn fuels. The energy to drive the turbine comes straight from the renewable resource. (Biofuels are the odd one out – they may be burned.)
- Energy from winds, waves and tides is available a lot of the time.
- The **kinetic energy** of winds and of moving water can be used to drive a turbine.

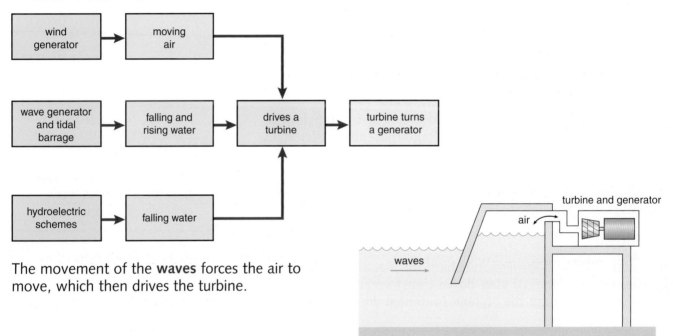

The movement of the **waves** forces the air to move, which then drives the turbine.

■ A wave-energy power station

The **tides** make water go in and out of an estuary. Turbines built into a **barrage** (a barrage is like a dam) across the estuary are driven by the moving water.

Hydroelectric systems trap water behind a dam. The dam is built across a river. When the trapped water is released, it falls down and drives the turbines.

gravitational potential energy → kinetic energy → electrical energy

■ A generator built into a tidal barrage

FOUNDATION

A hydroelectric **pumped storage** power station uses spare electricity to pump the water back up behind the dam. This makes the power station ready to generate again.

Other renewable energy resources

In some places, usually where there are volcanoes, the hot rocks below the Earth's surface turn water into steam. The steam that rises to the surface can be used to drive turbines. This is **geothermal energy**.

Solar cells transform the energy of the Sun's radiation into electricity.

Biofuels are things that grow. As they grow, they store chemical energy. Burning a biofuel transforms the chemical energy into heat, which is used to generate electricity in the same way as a coal-burning power station. Wood and straw are examples of biofuels.

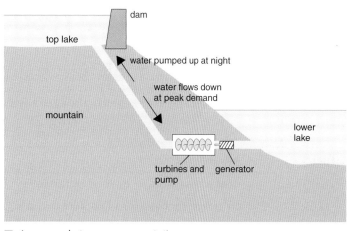

A pumped storage power station

check your understanding

19 What is the advantage of a tidal barrage compared with a wind generator? *(1 mark)*

  A It has no effect on the environment.
  B It generates electricity at predictable times.
  C It generates electricity only when it is windy.
  D It stores energy from surplus electricity.

20 Which type of power station uses the energy from hot rocks to produce steam? *(1 mark)*

  A a wind farm.
  B a wave-powered generator.
  C a hydroelectric system.
  D a geothermal power station.

21 What must be done to make sure wood is a renewable energy resource? *(1 mark)*

  A Chop as many trees down as possible.
  B Plant more slow-growing trees.
  C Plant one fast-growing tree for every tree chopped down.
  D Burn the trees as soon after chopping down as possible.

22 An area of solar cells 1 m^2 generates 12.5 W of power from 100 W of solar power.
  a) How many watts of electrical power would be generated from 400 W of solar power? *(2 marks)*
  b) What area of solar cells is needed to generate 1000 W of power? Assume the average solar power is 400 W per square metre. *(2 marks)*

exam tip

★ Remember to show your working in calculations such as question 22. Marks may be available for each stage of the calculation.

FOUNDATION

P1a Energy and electricity

Comparing energy resources

● Non-renewable fuels are reliable

If the power station has the fuel, electricity can always be generated.

Fuel	Advantages	Disadvantages
Coal and oil	Coal is easily moved by road or railway Oil is easily moved through pipes A small amount produces a lot of energy	Burning puts carbon dioxide and sulfur dioxide into the air Oil needs to be kept for other important industrial uses
Natural gas	Quick to start up and to switch off	Burning puts carbon dioxide into the air
Nuclear fuels	No polluting gases produced A very small amount produces huge amounts of energy	Some radioactive waste must be stored for thousands of years Serious accidents may spread radiation over large areas A nuclear power station is expensive to decommission (take to pieces) when it is not safe to use any more

● Renewable energy resources

- Produce no air pollution.
- The energy is free – **but** transforming the energy to electrical energy can be expensive.

Resource	Advantages	Disadvantages
Wind	Running costs are low Land around turbines can be used for farming	Some people think they spoil the view (visual pollution) and make unwanted noise Unreliable – do not generate all the time, only when the wind is strong enough Each turbine produces only a small amount of electricity, so a lot of turbines are needed
Waves	Running costs are low	Damaged by very rough seas
Tides	Reliable – tides happen twice a day, every day	Tidal barrages are expensive to build Barrage destroys the habitat of wading birds and other wildlife
Hydroelectric	Can be switched on and off quickly	Large areas of land may be flooded Flooding may destroy people's homes and affect plant and animal life
Solar	Ideal for remote places, or when only small amounts of electricity are needed	Do not work in the dark Large areas of solar cells needed to generate a lot of power
Geothermal	Massive amounts of energy available	Not easy to get the energy
Biomass	No extra carbon dioxide put into the atmosphere	A lot of land is needed to grow the crops

FOUNDATION

Deciding about building more power stations

- **Increased demand** for electricity means tough **decisions** have to be made.
- Fossil fuels will run out eventually.
- Many governments want to cut the amount of **carbon dioxide** put into the atmosphere.
- This means the amount of electricity generated from renewable energy resources needs to be increased.

Increasing renewable energy could be done by putting up more wind farms. But not everyone will want this. People on a remote, windy island may want a wind farm, but people living in an area of beautiful scenery may not want one built near them.

In the UK, more electricity could be generated from nuclear fuels and less from coal and gas. But how safe are nuclear power stations? And what happens to the nuclear waste? These questions need answers before any decision about nuclear energy is made.

exam tip

★ Read the information given in the question. This will help you give reasons **for** or **against** using a specific energy resource. If you are asked to give reasons for and against, and you only list reasons against, you will lose half your marks.

check your understanding

23. Which statement gives a reason for building new nuclear power stations? *(1 mark)*

    A Nuclear waste must be stored for thousands of years.
    B Nuclear fuel is non-renewable.
    C It costs a lot to take a power station to pieces when it is no longer safe.
    D A very small amount of fuel produces huge amounts of energy.

24. The doctor in a remote African village stores medicines in a freezer. Solar cells are used to generate the electricity needed to power the freezer. The main reason for using solar cells is that: *(1 mark)*

    A Solar cells generate electricity all of the time.
    B The village is a long way from any other electricity supply.
    C They are safer to use than a 230 V power supply.
    D A small number of solar cells generates a lot of electricity.

25. Which type of power station puts carbon dioxide into the air, but not sulfur dioxide? *(1 mark)*

26. Why doesn't a wind turbine produce electricity all the time? *(1 mark)*

FOUNDATION

P1b Radiation and the Universe

Waves and electromagnetic waves

● Waves move energy from one place to another

For example, if you throw a stone into a pond, ripples spread out from where the stone hits the water. The ripples move energy from one place to another.

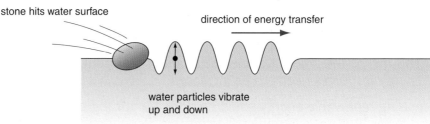

● Waves are described by their wavelength and frequency

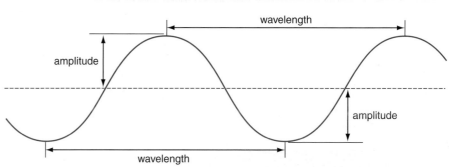

Wavelength = the distance from a point on one wave to the same point on the next wave.

Frequency = the number of waves produced each second.

The **wave equation** applies to all waves:

wave speed = frequency × wavelength

Units:

- wave speed is in **metres/second (m/s)**
- frequency is in **hertz (Hz)**;
- wavelength is in **metres (m)**;

> **exam tip**
> ★ If you need to work out wave speed, the equation will be given to you in the question. You don't need to remember or rearrange it. But you do need to remember the units.

● Electromagnetic radiation travels as waves

Electromagnetic waves form a continuous spectrum called the **electromagnetic spectrum**.

A **continuous spectrum** has no gaps in it. The visible light spectrum, which is a small part of the electromagnetic spectrum, is continuous; there are no gaps between the different colours.

There are seven types of electromagnetic wave in the electromagnetic spectrum. The waves are grouped together depending on how they are **reflected**, **absorbed** or **transmitted** by different substances and surfaces.

- Waves that hit a surface and bounce off have been **reflected**.
- Waves that go into a substance but do not come out have been **absorbed**.

> **exam tips**
> ★ Make sure you know the position of the waves in the spectrum.
> ★ If you have an exam question on this, always write about waves being reflected by a surface, **not** waves bouncing off a surface.

FOUNDATION

- Waves that go into a substance and come out have been **transmitted**. Light passes into and out of glass, so glass transmits light.

Types of electromagnetic wave

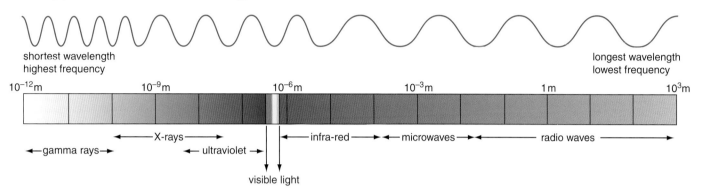

All electromagnetic waves:

- travel at the same speed through a vacuum;
- follow the wave equation for speed, frequency and wavelength;
- make substances that absorb them hotter.

Electromagnetic waves that are **absorbed by a metal** may produce a very small **alternating current (a.c.)** in the metal. The current will have the same **frequency** as the electromagnetic waves. This is how a TV aerial picks up a signal.

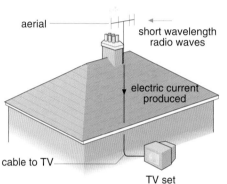

check your understanding

1. Radio waves with a wavelength of 100 000 m are used to communicate with a submarine. The waves are transmitted through the air at a frequency of 3000 Hz. Calculate the speed of the waves as they travel through the air and give the unit. *(3 marks)*

2. Which statement about electromagnetic waves is **not** true? *(1 mark)*

   A Radio waves travel through a vacuum at the same speed as light waves.
   B A substance that absorbs electromagnetic waves gets hotter.
   C Gamma rays do not follow the equation: wave speed = frequency × wavelength.
   D The frequency of an X-ray is measured in hertz.

3. Match each word (A–D) to one of numbers 1–4. *(4 marks)*

   A frequency   B seven   C two   D wavelength

   The electromagnetic spectrum is made up of ____1____ types of wave. ____2____ of these types of wave have a longer ____3____ and lower ____4____ than infra-red.

4. Which type of electromagnetic wave has a frequency between visible light and X-rays? *(1 mark)*

exam tips

★ If an exam question asks you to give your answer a unit, make sure that you do. The unit will be worth a mark.

FOUNDATION

P1b Radiation and the Universe

Uses of electromagnetic waves

What an electromagnetic wave is used for depends on how substances or surfaces **reflect**, **transmit** or **absorb** the wave.

● X-rays produce images on photographic film or computer screens

- X-rays go through flesh.
- Bone and metals absorb X-rays.

So bones and metal objects show up on an X-ray photograph or scan.

● Microwaves and infra-red are used for cooking

■ An X-ray photograph or scan shows an image of anything that absorbs X-rays

Microwaves go through substances such as plastic and glass. The containers used in microwave ovens are often plastic or glass.

Microwaves are absorbed by water molecules. Food contains water molecules.

In a microwave oven:

- the water molecules gain energy;
- the water heats up;
- the heat cooks the food.

Infra-red cooking works in a different way:

- warm objects emit infra-red;
- all objects absorb infra-red;
- the hot element of a toaster emits a lot of infra-red;
- a piece of bread absorbs the infra-red, becomes hot and is toasted.

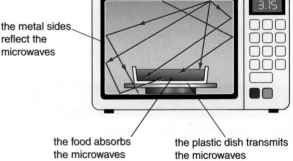

● Visible light, infra-red, microwaves and radio waves are used for communications

Visible light and infra-red signals can be sent down an **optical fibre**. The signal travels from one end of the fibre to the other by repeated reflections. The signal even follows any curves in the fibre.

Remote controls use infra-red to carry signals over short distances.

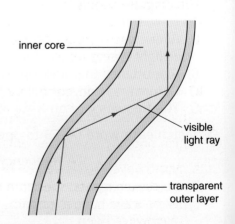

Microwaves are used:

- to send signals to and from **satellites** in space – this is because microwaves travel easily through the Earth's atmosphere;
- in **mobile phone** networks – the signals are transmitted over long distances from one tall aerial mast to another – if you phone another continent, the signal may go via a satellite orbiting in space.

Radio waves are transmitted for long distances by reflecting them off the **ionosphere** (a layer of ionised gas in the Earth's upper atmosphere).

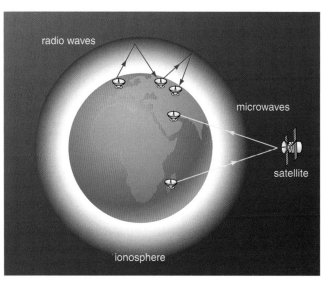

Communication signals can be analogue or digital

Analogue signals change continuously.

Digital signals are a series of 'on' and 'off' pulses. The pulses have only two values, which are often given the numbers 1 and 0.

Digital signals are better than analogue signals because:

- more information can be sent each second;
- received signals are higher quality (less **noise**);
- computers can be used to **process data** sent as a digital signal.

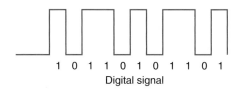

Digital signal

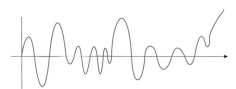

Analogue signal

check your understanding

5 Which part of the electromagnetic spectrum is used:
  a) to cook meat under a grill? (1 mark)
  b) in mobile phone networks? (1 mark)
  c) to send signals to a satellite? (1 mark)
  d) to obtain an image of a broken bone? (1 mark)

6 The diagram shows three different signals.
  a) Which signal (A, B or C) is digital? Give a reason for your answer. (1 mark)
  b) What name is given to a signal that is not digital? (1 mark)

7 Why are infra-red cameras able to detect people? (1 mark)

8 Which one of these statements is **not** true? (1 mark)

  A Microwaves are absorbed by water molecules.
  B Microwaves pass through a glass dish.
  C Microwaves are reflected by the metal mesh inside a microwave oven door.
  D Microwave ovens cook food by conduction.

exam tip

★ Make sure you know why digital signals are better than analogue signals.

FOUNDATION

P1b Radiation and the Universe

Hazards of electromagnetic waves

● X-rays and gamma rays are ionising radiations

Radiation that can remove electrons from atoms is called **ionising radiation**.

Most X-rays and gamma rays go straight through the human body, but some will always be **absorbed** by cells. The absorbed X-rays and gamma rays will **ionise** some atoms. This can cause cells to change (**mutate**) or can cause **cancer**.

Electromagnetic waves have different effects on living cells	
Type of wave	**Effect on living cells**
Gamma rays and X-rays	High doses kill cells Low doses can cause cells to mutate, or can cause cancer
Ultraviolet rays	Cause sunburn and suntan If absorbed by skin cells below the skin's surface, can cause skin cancer
Infra-red rays	Can cause burns
Microwaves	Heat the water in body cells – this may damage or kill the cells

● Reduce exposure to reduce the risk of cell damage

- **X-rays** – a small dose is not hazardous. But people who operate X-ray equipment could be exposed to large doses of X-rays, which is hazardous. So they often work behind concrete or lead screens. These materials are good absorbers of X-rays.
- **Ultraviolet** – causes skin cells to produce a brown substance called melanin. This is your suntan. Darker skin absorbs more UV, so less is absorbed by deeper layers of skin cells. This reduces the risk of skin cancer. Sun-block creams also reduce the amount of UV absorbed by the skin. They do this by reflecting or by absorbing the UV.
- **Microwaves** – mobile phone networks use microwaves. When absorbed, microwaves **heat** body cells because the cells contain water. But no-one is certain what the long-term effects of using a mobile phone or living near a phone mast may be.

FOUNDATION

The risks from mobile phones

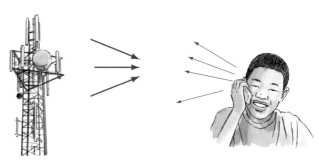

Recently, some scientists studied more than 4000 people and concluded that, in the first 10 years of using a mobile phone, there was no increased risk of cancer.

But **long-term risks** are still **unknown**. So experts advise mobile phone users to **reduce exposure** by keeping phone calls short.

check your understanding

9 It is hard for scientists to investigate the health of people who have been using mobile phones for more than 10 years. Which statement gives a reason why? *(1 mark)*

   A People cannot remember when they first started using a mobile phone.
   B Not many people have had a mobile phone for longer than 10 years.
   C Mobile phones do not work in some parts of the country.
   D People are not interested in the possible hazards of using a mobile phone.

> **exam tip**
> ★ Something about the possible dangers of using mobile phones is a likely exam question.

10 Which type of electromagnetic wave does not cause cancer? *(1 mark)*

   A microwaves   B infra-red   C gamma rays   D ultraviolet

11 The table gives the specific absorption rate (SAR) value for three different mobile phones. The higher the SAR value, the greater the amount of energy absorbed into the head when the phone is used. (SAR values are measured in a laboratory. The SAR value can be much lower when the phone is actually used.)

Mobile phone	SAR value
X	1.41
Y	0.69
Z	0.22

   a) Complete the following sentence: *(2 marks)*

   By measuring the SAR value in a ____1____, different phones can be tested under the ____2____ conditions.

   b) A phone sold in Europe cannot have an SAR value higher than 2.0. Which of the phones in the table could be sold in Europe? *(1 mark)*
   c) Does the SAR value prove a phone is safe to use? Give a reason for your answer. *(2 marks)*
   d) Some parents are going to buy a mobile phone for one of their children. Which phone (X, Y or Z) would you recommend they buy? Give a reason for your answer. *(2 marks)*

12 How does melanin help reduce the risk of skin cancer? *(2 marks)*

FOUNDATION

P1b Radiation and the Universe

Radioactivity

Particle	Charge
Proton	Positive
Neutron	Not charged
Electron	Negative

● An atom is made up of protons, neutrons and electrons

- Overall, atoms are electrically **neutral**:

  number of **protons** = number of **electrons**

- **Atoms** of the same element have the same number of **protons**.
- **Isotopes** of an element have different numbers of **neutrons**.

For example, there are three isotopes of carbon. The atoms all have the same number of protons but different numbers of neutrons.

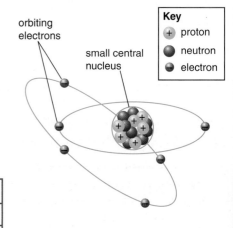

	Carbon-12	Carbon-13	Carbon-14
Number of protons	6	6	6
Number of neutrons	6	7	8

● Atoms of a radioactive substance have unstable nuclei

Radioactive substances emit **radiation** all the time. This radiation comes from the nucleus. How quickly the substance emits radiation (**decays**) does not change, no matter what is done to the substance.

● There are three types of nuclear radiation

These are alpha (α), beta (β) and gamma (γ) radiation – called nuclear radiation because all three come from the nucleus of an atom.

- An **alpha** (α) particle is the nucleus of a helium atom (two protons and two neutrons).
- A **beta** (β) particle is an electron.
- A **gamma** (γ) wave is a high-frequency electromagnetic wave.

exam tip

★ You need to remember these properties.

Properties of alpha, beta and gamma radiation				
Radiation	Ionising power	Range in air	Stopped by	Effect of electric and magnetic fields
Alpha (α)	Strong	A few centimetres	Thin paper	Very small deflection
Beta (β)	Moderate	A few metres	Few millimetres of aluminium	Large deflection (in the opposite direction to alpha)
Gamma (γ)	Weak	A few kilometres	Very thick lead	Not affected

FOUNDATION

Using nuclear radiation can be dangerous

Nuclear radiation can **ionise** atoms. This is what makes nuclear radiation dangerous to living cells (see page 98).

When the radiation is	Least dangerous	Most dangerous
Outside the body	Alpha – easily stopped by the air or by your skin	Gamma and beta – can pass through skin to damage cells and organs
Inside the body	Gamma and beta – can pass out of the body without being stopped	Alpha – easily ionise and damage cells

Ways of **reducing** the danger from nuclear radiation include:

- using thick sheets of dense material to absorb the radiation;
- wearing protective clothing;
- handling radioactive materials only with very long tongs, or better still remotely.

People working with radioactive substances may wear a badge containing photographic film. This does not stop them absorbing radiation, but how does show how much radiation they have been exposed to.

check your understanding

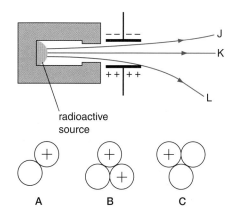

13 Which **one** of the following would not reduce the danger from the radiation emitted by a radioactive source? *(1 mark)*

  A Using long tongs to pick up a radioactive source.
  B Storing the radioactive source inside a box made from thick sheets of lead.
  C Making the radioactive source colder.
  D Wearing an apron lined with lead.

14 The diagram shows the paths taken through an electric field by the three different types of nuclear radiation.

  Which is the path of:
  a) the alpha particles? *(1 mark)*
  b) the beta particles? *(1 mark)*
  c) the gamma rays? *(1 mark)*

15 The diagram shows the nuclei of three atoms.

  Which two atoms are isotopes of the same element? Give a reason for your answer. *(2 marks)*

16 Why are high doses of nuclear radiation dangerous to people? *(1 mark)*

FOUNDATION

P1b Radiation and the Universe

Uses of radioactivity

Half-life

The average number of alpha, beta or gamma emissions in a certain time is called the **count rate**.

The amount of radiation emitted from a radioactive source goes down as time passes. This means that the count rate from a source always goes down. How quickly the count rate goes close to zero depends on the isotope. Some take millions of years, others only a few seconds. But the important thing is that all isotopes follow the same pattern.

The time it takes for the count rate or for the number of nuclei of a radioactive isotope to fall to half is called the **half-life**.

All radioactive isotopes have a half-life.

Radioactive wastes often have long half-lives. This means they decay slowly and will be radioactive for a long time. Some types of radioactive waste must be stored for thousands of years.

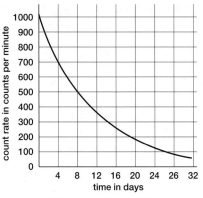

■ Iodine-131 has a half-life of 8 days

exam tips

★ You need to be able to find the half-life from a graph like this.

★ Make sure you look at the units on axes: time could be in seconds, minutes, days or years.

Uses of isotopes

What we use an isotope for depends on its half-life and the type of radiation emitted.

Isotope	Radiation emitted	Half-life
Cobalt-60	Gamma	5.3 years
Technetium-99	Gamma	6 hours
Manganese-52	Gamma	5.6 days
Strontium-90	Beta	28 years
Radon-220	Alpha	52 seconds

Medical tracers have a short half-life and usually emit gamma radiation.

A medical tracer is a radioactive isotope injected into a patient. The gamma rays (or sometimes beta particles) are detected outside the patient's body. Tracers have a short half-life, so the level of radiation inside the patient soon falls to a safe level (but is long enough for doctors to carry out a diagnosis). Technetium-99 is often used.

Tracers also have many industrial applications, including finding leaks in underground pipes.

Gamma rays kill bacteria – this means they can be used to:

- **sterilise** medical instruments;
- keep food fresh for longer.

exam tip

★ You need to think about the type of radiation emitted and the half-life.

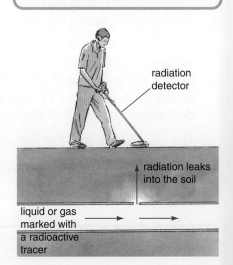

Cobalt-60 is often used to kill bacteria. The long half-life means the source does not need changing too often.

Isotopes used for quality control often have a long half-life.

A thick sheet of material absorbs more radiation than a thin sheet of material. So measuring the level of radiation passing through the sheet will automatically show if the thickness is changing. The control unit can then change the roller pressure to keep the sheet at the correct thickness.

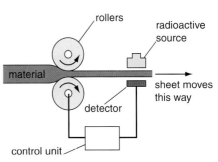

- For materials such as paper, cardboard and plastic, a beta source is used.
- For a metal sheet, a gamma source is used as beta particles would not get through at all.

check your understanding

17 Look at the list of isotopes in the table opposite. Which isotope would be used to:
  a) monitor the thickness of aluminium foil as it is produced? (1 mark)
  b) kill the cells in a cancerous tumour? (1 mark)
  c) follow oil flow in an underground pipe? (1 mark)

18 A radioactive isotope contains 100 000 undecayed nuclei. The isotope has a half-life of 4 days.
  a) How many undecayed nuclei will there be after 4 days? (1 mark)
  b) How many will there be after another 4 days? (1 mark)

19 The number of alpha, beta or gamma emissions from a source in one second is called the: (1 mark)

  A half-life.   B count rate.
  C rate of reaction. D decay constant.

20 The graphs show the decay of three different radioactive isotopes.

  Which isotope has:
  a) the longest half-life? (1 mark)
  b) the shortest half-life? (1 mark)
  c) the possibility of being used as a medical tracer? (1 mark)

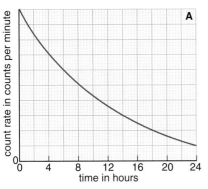
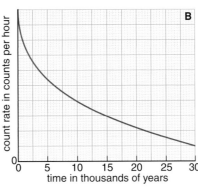
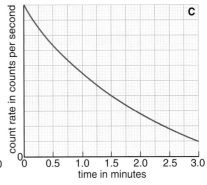

FOUNDATION

P1b Radiation and the Universe

Stars and telescopes

● **Telescopes are used to observe the Solar System and distant galaxies**

Stars emit visible light and other electromagnetic radiations such as radio waves or X-rays.

Optical telescopes detect visible light. They produce **magnified** images that we can see with our eyes. The telescopes are often built on **high mountains**, where the atmosphere is thinner. The thinner the atmosphere, the less it **distorts** the light from a star or galaxy, and so the clearer the image. Weather conditions also affect the quality of the image – **clouds** block light.

Radio telescopes detect the weak radio signals given off by distant stars and galaxies. The signals are processed by computers to produce an image of the stars and galaxies emitting the signals.

Other types of telescope detect gamma rays, X-rays, ultraviolet or infra-red.

■ Telescopes are placed on mountain-tops as the layer of atmosphere above the telescope is thinner

Some telescopes are on satellites orbiting the Earth

Space telescopes detect different types of electromagnetic radiation. The Hubble space telescope detects infra-red, visible light and ultra-violet radiations.

	Advantages	Disadvantages
Telescopes on Earth	Easy to maintain and repair Telescopes can be linked to give greater detail	Need huge structures to hold and move the mirror Images distorted by the atmosphere Light absorbed by clouds
Telescopes in space	Images not distorted by the atmosphere Signals not affected by the weather Can detect objects further away in space	Very expensive to build and put into space Difficult to maintain or repair

check your understanding

21. The following two statements apply to one particular telescope:

    It detects weak radio signals.

    It is easy to repair and maintain.

    a) Is the telescope on Earth or in space? (1 mark)
    b) What type of telescope is it? (1 mark)

22. Which of the following statements gives the reason why Earth-based telescopes are often built on high mountains? (1 mark)

    A There is less atmospheric distortion.
    B They produce less visual pollution.
    C Few people live high on a mountain.
    D The land is not used for farming.

FOUNDATION

P1b   Radiation and the Universe

Expanding Universe and 'big bang'

● The 'big bang' theory predicts that the Universe is expanding!

The **'big bang' theory** suggests that the Universe was once concentrated at a tiny point in space. There was a huge explosion, causing everything to expand outwards into the Universe we have today. This is only a theory – scientists cannot prove the theory, but they can look for evidence to support it.

● What is red-shift?

If a light source moved away from you quickly enough, the light waves you see would also have a longer wavelength. This would make the light appear as if it has moved to the red end of the spectrum. The light would show a **red-shift**.

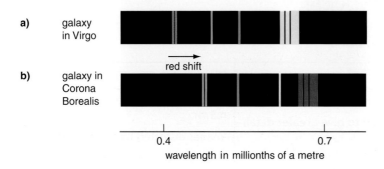

106   FOUNDATION

Red-shift provides evidence for an expanding Universe

When scientists look at the light from distant **galaxies**, it shows a red-shift. It is called this because the light looks as though it has moved towards the red end of the spectrum. This happens because the galaxies are **moving away** from the Earth.

Scientists have also observed that the further a galaxy is from the Earth, the bigger the red-shift. This happens because the galaxies furthest away are moving the fastest.

So at some time in the past, the galaxies must have been closer together. Red shift provides the evidence to support the 'big bang' theory and the idea of an expanding Universe.

exam tip

★ You must be able to describe red-shift and say why it is evidence for an expanding Universe and the 'big bang' theory.

check your understanding

23 Complete the following sentences. *(4 marks)*

Light from distant galaxies shows a ___1___, so galaxies are moving ___2___ from the Earth. This means the Universe is ___3___, something predicted by the ___4___ theory.

24 Which one of the following statements about the 'big bang' theory is correct? *(1 mark)*

A It is the only way of explaining the origin of the Universe.
B There is scientific evidence to support the theory.
C It is based on scientific and religious facts.
D Scientists have proof that it happened.

Answers

B1a

Nervous system and reflexes

1 (12 marks)

Sense	Sense organ	Receptor cells sensitive to
sight	the eye	light
hearing	ears	sound
taste	tongue	chemicals
smell	nose	chemicals
touch	skin	pressure/temperature/pain
balance	ears	changes in position

2 a) Transmit information/impulse (*1 mark*) from receptor to brain (*1 mark*).
b) Sensory nerves take nerve impulses into the spinal cord; motor nerves take nerve impulses away from the spinal cord. (*2 marks*)

3 1, stimulus; 2, sensory; 3, brain (*3 marks*)

Hormones, the menstrual cycle and fertility

4 D (*1 mark*)

5 B (*1 mark*)

6 1, hormone; 2, blood; 3, target organ (*3 marks*)

7 May be stressful, expensive, multiple birth, ethical issues. (*4 marks*)

Controlling internal conditions

8 a) To provide the cells with energy. (*1 mark*)
b) Sweat, urinating, by breathing out (also faeces). (*3 marks*)

9 1, water; 2, dehydrated; 3, sodium (*3 marks*)

10 B (*1 mark*)

Drugs: good and bad

11 a) Become dependent (*1 mark*), suffer withdrawal symptoms without (*1 mark*).
b) Cocaine, heroin. (*2 marks*)
c) Nicotine, alcohol. (*2 marks*)

12 1, side effects; 2, human tissue; 3, live animals; 4, clinical trials (*4 marks*)

13 C (*1 mark*)

Tobacco and cannabis

14 a) Nicotine (*1 mark*) is addictive. (*1 mark*)
b) Awareness of harm to health/link to lung cancer. (*1 mark*)
c) Better treatment available, or earlier diagnosis. (*1 mark*)
d) Less chance of low-birth-weight babies. (*1 mark*)

15 a) No, the correlation may be due to chance or other causes. (*2 marks*)
b) Long sunny days. (*1 mark*)

Diet and exercise

16 1, energy requirements; 2, glucose; 3, metabolic rate (*3 marks*)

17 C (*1 mark*)

18 Malnourished means wrong balance of nutrients; undernourished means total nutrients is not enough. (*2 marks*)

19 The warmer it is, the less food you need to keep warm. (*1 mark*)

20 a) Swimming. (*1 mark*)
b) Total amount of activity uses more energy. (*1 mark*)

Problems with 'bad' diets

21 1, obese; 2, saturated fat; 3, heart disease; 4, statins; 5, side effects (*5 marks*)

22 Any two from: obesity; high cholesterol; smoking, genetics; high fat diet; stress. (*2 marks*)

23 a) Any two foods high in saturated fat, e.g. dairy and meat products. (*2 marks*)
b) Narrows arteries. (*1 mark*)
c) Replace saturated fat with unsaturated fat. (*1 mark*)

Fighting disease

24 a) Microorganism that causes disease. (*2 marks*)
b) Bacteria and viruses. (*2 marks*)

25 Produce antibodies; produce antitoxins; engulf and destroy microorganisms. (*3 marks*)

26 a) Dead or inactive forms of the pathogen. (*1 mark*)
b) Measles, mumps, rubella/German measles. (*3 marks*)
c) Risk of side effects caused by vaccine. (*1 mark*)

27 C (*1 mark*)

The fight against disease – then and now

28. a) Antibiotics do not kill viruses, only bacteria. (*1 mark*)
    b) Antibiotic-resistant bacteria. (*1 mark*)
    c) So antibiotic-resistant bacteria do not develop as quickly. (*1 mark*)
29. C (*1 mark*)
30. B (*1 mark*)
31. The microorganism that causes MRSA is transferred by contact/touch (*1 mark*), good hygiene reduces spread of infections from doctor/nurse/visitor to patient/stops transfer of microorganism (*1 mark*).

B1b

Adapt and survive

1. Any two from: toxins; colouration; spines; bright armour. (*2 marks*)
2. C (*1 mark*)
3. Reduced surface area (*1 mark*) so less heat loss (*1 mark*) – no marks for just 'to keep warm'.

Populations and competition

4. a) Habitat destruction, hunters, egg collectors, pollution. (*1 mark*)
   b) Protection of habitat, re-introduction into wild, more food, less pollution. (*2 marks*)
5. a) The rats have enough food to survive and reproduce (*1 mark*) and there are no predators to eat the offspring (*1 mark*).
   b) The rats compete with each other for resources. (*2 marks*)
6. B (*1 mark*)

Variation and inheritance

7. Cell before nucleus (*1 mark*), chromosome before gene (*1 mark*).
8. C (*1 mark*)
9. D (*1 mark*)

Reproduction and cloning

10. Sexual reproduction involves two parents (male and female), asexual only one parent. (*1 mark*)
11. Sexual reproduction as it mixes genes from two parents (*1 mark*); asexual reproduction produces genetically identical offspring/clones (*1 mark*).
12. Genetically identical to parent (*1 mark*).
13. Any two of: twins; cuttings or runners from plants; tissue culture. (*2 marks*)
14. C (*1 mark*)

Genetic engineering

15. Can apply weedkiller over a whole field and not worry about killing the crops. (*1 mark*)
16. 1, genes; 2, chromosome; 3, enzymes (*3 marks*)
17. C (*1 mark*)

The fossil record

18. D (*1 mark*)
19. C (*1 mark*)

How evolution happens

20. Any example of a skill or characteristic acquired by a parent that is not passed to offspring (e.g. learning Chinese, bodybuilder's large muscles, losing a limb in an accident). (*1 mark*)
21. 1, Lamarck; 2, Darwin; 3, acquired; 4, inherited (*4 marks*)

How do humans affect the environment?

22. Any metal or ore; any fossil fuel. (*2 marks*)
23. a) The number/amount of lichen (*1 mark*) indicates whether pollution is present (*1 mark*).
    b) There would be less lichen near the road, as there is more air pollution there.
    c) There could be random variations over the site, so one sample would not be representative (*1 mark*); to improve reliability of data (*1 mark*).
24. 1, acid rain; 2, global warming; 3, sewage; 4, deforestation (*4 marks*)

Global warming

25. More cattle; more rice cultivation.
26. Concentration increases (*1 mark*), as there are fewer trees to remove carbon dioxide from the atmosphere (*1 mark*) (note: burning is not mentioned in the question so cannot say more carbon dioxide is released).
27. D (*1 mark*)

Sustainable development

28. Recycling paper means fewer forests are cut down; recycling glass and metals means that fewer quarries are dug. (*2 marks*)
29. A (*1 mark*)

C1a

Atoms, elements and the Periodic Table

1. As on page 40. (*3 marks*)
2. An element – it is made of only one sort of atom. (*1 mark*)
3. C (*1 mark*)
4. C (*1 mark*)

Reactions, formulae and balanced equations

5. 1, bonds; 2, electrons; 3, compound. (*3 marks*)
6. D (*1 mark*)

Products from limestone

7. copper carbonate → copper oxide + carbon dioxide. (*3 marks*)
8. C (*1 mark*)
9. C (*1 mark*)

Extracting metals

10. The least reactive metals, such as gold, platinum and silver. (*1 mark*)
11. A (*1 mark*)
12. A (*1 mark*)

Mining, quarries and recycling

13. Any three benefits such as: the quarry provides jobs; raw materials for building; new road links may be built. (*3 marks*)
14. Any three drawbacks such as: it is ugly while it is being dug; there is more noise/dust pollution than before; wildlife habitats are destroyed. (*3 marks*)
15. C (*1 mark*)

Using metals and alloys

16. Make it brittle. (*1 mark*)
17. B (*1 mark*)
18. D (*1 mark*)

Crude oil and fuels

19. D (*1 mark*)
20. D (*1 mark*)
21. B (*1 mark*)
22. Colour gets darker with molecule size. (*1 mark*)

Burning fuels

23. Soot/particles (*1 mark*) blocks sunlight (*1 mark*) so decreases temperature (global dimming) (*1 mark*).
24. A (*1 mark*)
25. D (*1 mark*)

Cleaner fuels

26. D (*1 mark*)
27. Advantages: renewable (do not use up non-renewable fuels); clean (no pollutants when burnt) / disadvantages: takes a lot of land to grow; less energy output. (*2 marks*)
28. ethanol + oxygen → carbon dioxide + water. (*2 marks*)

C1b

Cracking crude oil

1. C (*1 mark*)
2. D (*1 mark*)
3. A (*1 mark*)

Making ethanol

4. It only produces water (*1 mark*) and carbon dioxide (*1 mark*) only (*1 mark*).
5. A (*1 mark*)
6. C (*1 mark*)

Making polymers

7. poly(propene) (*1 mark*)
8. 1, alkenes; 2, monomers; 3, large; 4, ethene (*1 mark each*).
9. C (*1 mark*)

Waste-disposal problems

10. B (*1 mark*)
11. Any four valid points, e.g.: (*4 marks*)

	Advantages	Disadvantages
Crude oil products as fuels	High energy output Easy to produce	Produces pollution Uses finite resource only once
Crude oil products as raw materials	Can make much use of the oil Many products Many uses	Reduces availability of fuel

Vegetable oils and fuels

⑫ a) Olive oil. (1 mark)
  b) Less risk of heart disease and stroke (1 mark) due to less cholesterol (1 mark).
  c) React them with hydrogen (1 mark) in the presence of a nickel catalyst (1 mark) at about 60 °C (1 mark).
⑬ Crops (such as sugar cane). (1 mark)
⑭ D (1 mark)
⑮ D (1 mark)

Food additives and emulsifiers

⑯ A (1 mark)
⑰ A (1 mark)
⑱ B (1 mark)

The Earth and continental drift

⑲ as on page 70. (4 marks)
⑳ C (1 mark)
㉑ support (1 mark)
㉒ B (1 mark)

Plate tectonics

㉓ A (1 mark)
㉔ B (1 mark)
㉕ C (1 mark)
㉖ Earthquakes occur at the boundaries between tectonic (1 mark) plates as the result of sudden movement (1 mark).

Gases in the atmosphere

㉗ D (1 mark)
㉘ C (1 mark)
㉙ A (1 mark)

Theories about the atmosphere

㉚ C (1 mark)
㉛ D (1 mark)
㉜ B and D (2 marks)

P1a

Thermal radiation

① 1 B; 2 D; 3 A; 4 C (4 marks)
② A, matt white; B, shiny black; C, matt black; D, shiny silver (4 marks)
③ a) Y; b) Y; c) 22 °C (3 marks)

Conduction and convection

④ B and C (2 marks)
⑤ D (1 mark)
⑥ Colder, heavier air falls keeping the rest of the fridge cold. (2 marks)
⑦ 1 conductors; 2 insulators; 3 particles; 4 liquid (4 marks)

Reducing rates of heat transfer

⑧ a) conductor;
  b) increases;
  c) radiation;
  d) air (4 marks)
⑨ B (1 mark)

Energy efficiency

⑩ Carpet, 20 years; double glazing, 40 years; hot water tank, 5 years. (3 marks)
⑪ a) 2300 (1 mark)
  b) D (1 mark)
⑫ C (1 mark)

Electrical power and energy costs

⑬ B (1 mark)
⑭ D (1 mark)
⑮ £11.40 (3 marks)

Generating electricity and the National Grid

⑯ Natural gas. (1 mark)
⑰ Nuclear fuel is not burned / heat produced by nuclear fission. (1 mark)
⑱ a) 1250 MW (1 mark)
  b) 1¼ years / 1 year and 3 months (1 mark)

Renewable energy resources

⑲ B (1 mark)
⑳ D (1 mark)
㉑ C (1 mark)
㉒ a) 50; b) 20 (4 marks)

FOUNDATION 111

Comparing energy resources
- 23. D (*1 mark*)
- 24. B (*1 mark*)
- 25. natural gas. (*1 mark*)
- 26. not always windy. (*1 mark*)

P1b

Waves and electromagnetic waves
1. 300 000 000 m/s (*3 marks*)
2. C (*1 mark*)
3. 1 seven; 2 two; 3 wavelength; 4 frequency (*4 marks*)
4. ultraviolet (*1 mark*)

Uses of electromagnetic waves
5. a) infra-red;
   b) microwaves;
   c) microwaves;
   d) X-rays. (*4 marks*)
6. a) B; The signal shows only two values, none in between. (*1 mark*)
   b) analogue. (*1 mark*)
7. people give out (emit) infra-red. (*1 mark*)
8. D (*1 mark*)

Hazards of electromagnetic waves
9. B (*1 mark*)
10. B (*1 mark*)
11. a) 1, laboratory; 2, same (*2 marks*)
    b) X, Y, Z (*1 mark*)
    c) No: there is no proof that the absorbed energy is totally harmless. (*2 marks*)
    d) Z, the amount of radiation absorbed is the lowest so can expect it to be the safest. (*2 marks*)
12. Absorbs UV reducing the amount that reaches deeper layers of skin. (*2 marks*)

Radioactivity
13. C (*1 mark*)
14. a) alpha, J;
    b) beta, L;
    c) gamma, K (*3 marks*)
15. A and C, have same number of protons. (*2 marks*)
16. cause damage to the cells, or cancer. (*1 mark*)

Uses of radioactivity
17. a) strontium-90;
    b) cobalt-60;
    c) manganese-52 (*3 marks*)
18. a) 50 000; b) 25 000 (*2 marks*)
19. B (*1 mark*)
20. a) C; b) B; c) C (*3 marks*)

Stars and telescopes
21. a) Earth;
    b) radio telescope (*2 marks*)
22. A (*1 mark*)

Expanding Universe and 'big bang'
23. 1, red-shift; 2, away; 3, expanding; 4, big bang. (*4 marks*)
24. B (*1 mark*)

112 FOUNDATION

Index

absorb (radiation) 78, 95, 96, 98
accuracy of measurements 79
acid rain 34, 54, 56
adaptation 20, 32
addiction 8
addictive drugs 10
adult cell cloning 27
aerial 95
air pollution 34
alcohol 8
alkanes 52, 58
alkenes 58, 59, 62
alloy 50
alpha particle 100
alternating current 95
aluminium, extraction 46
aluminium, uses 51
analogue signal 97
animal testing 9
anomalous result 35, 78
antibiotic-resistant bacteria 18, 19
antibiotics 18, 19
antibodies 17
antigens 16
antiseptic 18
antitoxins 17
argon, uses of 75
arteries 14
arthritis 14
artificial colours 68
asexual reproduction 26
atmosphere, gases in 74, 76–77
atmosphere, theories of formation 76
atoms 40, 43
autism 17
avian flu 19

bacteria 16, 18
balanced argument 56
balanced diet 12
balancing chemical equations 43
bias in data 7, 35
'big bang' theory 106
biodegradable 64
biodiesel 67
biofuels 56, 60, 91
blast furnace 46, 50
blood pressure 14
blood sugar 6
bonds 42

bromine water 59, 66
building materials 44

cacti 20
camel 21
camouflage 20
cancer 98
cannabis 10, 11
carbohydrates 12
carbon dioxide in the atmosphere 34, 36, 74, 75, 77, 92
carbon dioxide, test for 54
carbon monoxide 10, 54
carbon-carbon bonds 52, 58
carbon-neutral 67
carcinogenic 10
cardiovascular disease 14
cast iron 50
catalyst 58, 60
causal link 10, 11
cement 44
chemical bonds 42
chemical formulae 42
chemical symbols 40, 42
cholesterol 14, 15, 67
chromatography 68
chromosomes 25
climate change 30
clinical trials 9
cloning 26, 27
cocaine 8
combustion 36, 54
combustion, products of 54
comets 76
common ancestor 31
competition 22, 30, 32
compounds 40, 46
concrete 44, 45
condense 53
conduction 80, 82
conductor 80
conservation of mass 43
conserving resources 38
continental drift 70, 71
contraceptives 5
control variable 7, 9, 11, 15, 82
convection 72, 80, 82
coordination 2
copper, extraction 46, 47
copper, uses 47
core 70

coronary heart disease 14
correlation 10, 37, 53, 75
corrosion 51
cost effective 85
cost of electricity 86
count rate 102
cracking 58, 60
credible evidence 10
crude oil 52, 58, 63
crust 70, 72
cuttings 26

dangers of radioactivity 101
Darwin, Charles 32, 33
deficiency diseases 12
deflection of radiation 100
deforestation 34, 36, 38
diabetes 6, 14
diet 12, 14
digital signal 97
distillation 52, 56, 66
DNA 25, 28
Doppler effect 106
double bonds 52, 58
double glazing 83
drugs 8, 9, 11, 15

E-numbers 68
Earth, structure of 70
earthquakes 73
economic value of ores 46
effector 2, 6
efficiency, energy 84
egg cell 4, 5, 26
electrical energy 86, 89
electricity demand 92
electricity meter 86
electrolysis 46
electrolytes 6
electromagnetic radiation 78
electromagnetic spectrum 95
electrons 40, 100
electrons, sharing and transferring 42
element 40
embryo transfer 27
emit (radiation) 78
emulsifiers 68
emulsions 68
endangered species 38
energy efficiency 49, 84

FOUNDATION 113

Index

energy in food 12
energy in fuels 57
energy resources 34, 38, 67, 90, 92
energy transfers and transformations 84, 86
energy-efficient lamp 85
environmental factors 24
enzyme 28
epidemic 18
equations 43
errors in measurements 35, 78, 79
ethanol as a fuel 56, 60
ethene 60
evaluating benefits and drawbacks 29, 67, 69
evaluating claims 7, 12, 64
evaluating data and measurements 7, 11, 78
evolution 30, 31, 32
evolution, theories of 33
exercise 12, 13
expanding universe 106
exposure to radiation 99
extinct 30
extracting metals 46
extrapolate 79

fair test 82
fats 12, 15, 66, 67
fermentation 60
fertilisation 26
fertility treatment 5
fibre 12
flammability 53
flavourings 68
flue gas desulfurisation 56
follicle-stimulating hormone (FSH) 4
food additives 68
fossil fuels 34, 36, 55, 75, 77
fossil record 31
fossils 70
fractional distillation 52, 58, 60
fractions 52
frequency (waves) 94
fuels 53, 54, 55, 56, 58, 66
fuels, alternative 56
fusion cell cloning 27

galaxies 104, 106
gametes 26
gamma rays 98, 100, 102
generating electricity 88

genes 24, 25, 28
genetic engineering 28
geothermal energy 91, 93
glands 4
glass 44, 45
global dimming 54
global warming 34, 36, 54
global warming, impacts of 37
glucose 6, 7
GM crops 28, 29
greenhouse effect 36
groups of elements 41

habitat 22
half life 102, 103
hazards of electromagnetic waves 98
heart disease 14
heat energy 78, 80
heat transfer, reducing 82
helium, uses 74
herbicide 34
heroin 8
high blood pressure 14
high density lipoproteins (HDLs) 15
high-carbon steel 50
homeostasis 6
hormones 4
hot-water systems 81
human impact on the environment 23, 34
hydrocarbons 52
hydroelectric systems 90, 93
hydrogels 62
hydrogen fuel 56
hydrogenation 66

immune system 17
independent variable 82
indicator species 35
infection, resistance to 10
infectious diseases 16, 18
infertility 5
infra-red radiation 78, 97, 98
ingest 17
inherited factors 13, 24
insulation 83
insulator 80
invertebrates 35
iodine solution 59, 66
ion content of the body 6
ionise 101
ionising radiation 98
ionosphere 97

iron, extraction 46, 50
isotope 100
IVF 5

joules 84

kilowatt 86
kilowatt-hour 86
Kyoto Agreement 38

Lamarck, Jean-Baptiste 33
landfill sites 49, 64
lichen 35
life, beginnings of 30
limestone 44, 45
limewater 54
lipoproteins 15
liver 14, 15
liver damage 8
low density lipoproteins (LDLs) 15
low-carbon steel 50
lung cancer 10
luteinising hormone (LH) 4

malnourished 12
mantle 70, 72
margarine 66, 67
matt surfaces 78
medicines 8
medicines, traditional 9
menstrual cycle 4
metabolic rate 12, 13
metals, properties 51
methane 36
microorganisms 16
microwaves 96, 97, 98
minerals 12, 46
mining 46, 47
mixtures 50, 52, 68
MMR vaccine 17
mobile phone communication 97
mobile phones, risk 99
models 37
molecules 42
mortar 44
motor neurones 3
MRSA (methicillin-resistant *Staphylococcus aureus*) 18
mutation 18, 32, 98

National Grid 89
natural selection 18, 32
nerves 2, 3
neurones 2

neutron 100
nicotine 10
noble gases 74
noise in signals 97
non-biodegradable 64
non-renewable 34, 38, 56, 88
nuclear fission 88
nuclear power 88, 92
nuclear radiation 100, 101
nucleus 25, 40

obesity 14, 67
oestrogen 5
opinions, evaluating 11, 48, 55
optical fibre 97
ores 46
ovaries 4
overweight 13, 14
ovulation 4

painkiller 16
pancreas 6
pandemic 18
particles from combustion 54, 60
pathogens 16, 22
payback time 85
Periodic Table 41
pesticide 34
photosynthesis 36, 77
pituitary gland 4
placebo 9
plastics 58, 62
plate tectonics 72
pollution 23, 34–35, 38, 92
poly(chloroethene) 62
poly(ethene) 62
poly(propene) 62
polymers 62, 65
polymers, disposal 64
population growth 34
potential difference 89
power 86
power stations 88, 92
power transmission 89
precision of measurements 79
predator 22, 30
preservatives 68
prey 22
processed food 15
protein 12
proton 100
PVC 62

quadrat 35

quarrying 44, 48
quicklime 44

radiation from radioactive substances 100, 101
radiation of heat energy 78, 82
radiators 81
radio telescopes 104
radio waves 95, 97
radioactive decay 100
radioactive waste 92, 102
radioactivity 100
radioactivity, uses of 102
random error 79
random sample 35
reaction time 8
reactivity series 46
receptor cells 2, 6
recycling 38, 49, 64
red-shift 106
reduction 46
reflect 78, 95, 96, 97
reflex 3
relay neurones 3
reliable data 7, 9, 15, 78
renewable energy resources 67, 90, 92
repeated measurements 78
representative sample 35
resources, limited 22, 49
respiration 13

salt 14, 15
sample size 10, 15
sampling technique 35
satellite communication 97
saturated fat 15, 66, 67
saturated hydrocarbons 52
sea level rise 37
Semmelweiss, Ignaz 18
sense organs 2
sensitivity of measurements 79
sensory neurones 3
sewage 34
sexual reproduction 26
shape-memory alloys 51
shape-memory polymers 62
shiny surfaces 78, 83
side effects 5, 9
skin cancer 98
slaked lime 44
slime 62
smart alloys 51
smart materials 44

smart polymers 62
smoking 10, 11
solar cells 91, 93
soot 54
sperm 26
sports drinks 7
stainless steel 50
stars 104
start-up time 89, 92
statins 15
steel 50
stimulus 2
sulfur dioxide 34, 54, 56
surface area-to-volume ratio 20
surveys 10, 15
survival advantage 32
sustainable development 38
synapse 3
systematic error 35

tar 10
target cells 4
tectonic plates 72
telescope 104
temperature control in body 6
temperature, measuring 78
thalidomide 9
thermal decomposition 45, 58
thermal radiation 78
thermoplastic 62
thermosetting 62
thorns 20, 21
tidal power 90, 93
tissue culture 27
titanium, uses 51
tobacco 10
toxins 16
tracers 102
transferring energy 84
transformer 89
transforming energy 84
transition metals 50
transmit (radiation) 95, 96
transmitter chemicals 3
turbine 88
twins 26

ultraviolet radiation 98
Universe, expansion of 106
unsaturated fats 15
unsaturated hydrocarbons 58
unsaturated oils 66
unsaturation, test for 66
useful energy 84

Index

uterus 4

vaccine 17
vacuum flask 82
valid data 7, 10
variation 24, 32
vegetable oils 66, 68
virus 16
viscosity 53, 62

vitamins 12
volcanoes 73, 76

Warfarin 32
warning colours 21
waste disposal 64
wasted energy 84, 89
water content of the body 6
watt 86

wave power 90, 93
wavelength 94
waves 94–95
Wegener, Alfred 70
white blood cells 17
wind power 90, 92, 93
withdrawal symptoms 8

X-rays 96, 98

D2919

university college
for the creative arts

581 HUNT

WITHDRAWN
FROM STOCK